Springer Japan KK

A. Kitabatake
I. Sakuma (Eds.)

Recent Advances in Nitric Oxide Research

With 38 Figures

Springer

Akira Kitabatake, M.D.
Department of Cardiovascular Medicine
Hokkaido University School of Medicine
N-15, W-7, Kita-ku, Sapporo 060-8638, Japan

Ichiro Sakuma, M.D., Ph.D.
Department of Cardiovascular Medicine
Hokkaido University School of Medicine
N-15, W-7, Kita-ku, Sapporo 060-8638, Japan

ISBN 978-4-431-68010-9 ISBN 978-4-431-67929-5 (eBook)
DOI 10.1007/978-4-431-67929-5

Printed on acid-free paper

Originally published by Springer-Verlag Tokyo Berlin Heidelberg New York in 1999

Typesetting: Camera ready copy from Editors

SPIN: 10677273

Preface

The Sapporo International Symposium on "Recent Advances in Nitric Oxide Research" was held in Sapporo, Japan, in 1997, following the Fifth International Meeting on the Biology of Nitric Oxide in Kyoto, Japan, organized by Dr. Salvador Moncada, Dr. Noboru Toda, and Dr. Hiroshi Maeda. The field of nitric oxide research continues to expand rapidly, and our understanding of the physiological and pathophysiological roles of NO has increased greatly. The Kyoto Meeting was stimulating and informative, providing impetus for the Sapporo Symposium, which I had the great honor to organize.

To communicate the information from these events, Dr. Ichiro Sakuma and I decided to publish this book. The contents of its chapters were contributed by the participants who were active at the Sapporo symposium and cover the majority of the presentations made during that symposium. Dr. Csaba Szabo of Children's Hospital Medical Center in Cincinnati (U.S.A.) reviews the roles of peroxynitrite and poly(ADP-ribose)synthetase in shock, inflammation, and reperfusion injury, and Dr. David A. Geller and his colleagues of the University of Pittsburgh (U.S.A.) review the regulation and function of NO in the liver. As contributions from the Hokkaido University School of Medicine (Sapporo), Dr. Hiroko Togashi and colleagues present their data on transient cerebral ischemia and NO production, Dr. Ichizo Tsujino and colleagues describe measurement of exhaled NO from the lung, and my colleagues and I demonstrate a new aspect of NO versus hemoglobin interaction. Dr. Nobuyuki Ura and colleagues of the Sapporo Medical University School of Medicine (Sapporo) deal with an important subject, the role of renal NO and kinins on renal water-sodium metabolism. Dr. Yoshiyuki Hattori of Dokkyo University (Mibu) presents his recent results with tetrahydrobiopterin synthesis in cardiac myocytes, which he has studied with Dr. Steven Gross of Cornell University Medical College (U.S.A.). Dr. Gautum Chaudhuri of the University of California, Los Angeles (UCLA)(U.S.A.) reviews the contribution of NO in ovarian cancer, and Dr. Toshio Hayashi of Nagoya University (Nagoya), who studied at UCLA, writes about estrogen and NO.

The result is a book that I believe will provide valuable, up-to-date sources of information to the reader in various fields. While this book was in the process of being edited, the 1998 Nobel prize in medicine was awarded for work on nitric oxide. I am delighted to say that two authors of this book, Dr. Gautum Chaudhuri and Dr. Toshio Hayashi, worked with Dr. Louis Ignarro, one of three 1998 Nobel laureates in medicine.

I am deeply indebted to all of the authors for their time and effort despite their tight schedules. I would also like to thank Dr. Ichiro Sakuma, Hokkaido University, and the editorial staff of Springer-Verlag, Tokyo, for their constant support in the publication of this book.

Akira Kitabatake
November, 1998

Contents

Part 1

NO and Circulation

Role of peroxynitrite and poly-(ADP-ribose) synthetase in shock, inflammation, and ischemia-reperfusion injury

Csaba Szabó[1]

SUMMARY. Peroxynitrite is a reactive oxidant produced from the reaction of nitric oxide (NO) and superoxide. Immunohistochemical and biochemical evidence demonstrate the production of peroxynitrite in endotoxic and hemorrhagic shock, bowel inflammation, allergic encephalomyelitis, and in various forms of ischemia - reperfusion injury, including myocardial and splanchnic reperfusion and stroke. Peroxynitrite induces tyrosine nitration, lipid peroxidation, inhibition of respiratory enzymes and inactivation of membrane channels. Peroxynitrite also triggers DNA strand breakage, with subsequent activation of the nuclear enzyme poly-ADP ribosyl synthetase (PARS), with consequent energy depletion of the cells. Experimental evidence, using pharmacological inhibitors or PARS and/or genetically engineered cells or animals lacking PARS enzyme demonstrates the role of the PARS pathway in the cellular injury associated with stroke, myocardial and splanchnic ischemia-reperfusion, shock and multiple organ failure, diabetes and arthritis.

KEY WORDS: Inflammation, nitric oxide, superoxide, shock, DNA strand breaks, arthritis, stroke, heart

---

[1] Children's Hospital Medical Center, Division of Critical Care, 3333 Burnet Ave, Cincinnati, Ohio 45229, USA

Generation and actions of peroxynitrite

Sources of peroxynitrite

Nitric oxide and superoxide rapidly react to yield the toxic reaction product, peroxynitrite anion ($ONOO^-$) [1-3]. The sources of superoxide include the mitochondrial chain, NADPH oxidase, xanthine oxidase, lipid peroxidation, and other sources [4-7]. The sources of NO for the generation of peroxynitrite are the three isoforms of NO synthase: the endothelial isoform (ecNOS), the brain isoform (bNOS) and the inducible isoform (iNOS) [8]. Under low cellular arginine concentrations, NOS produces both NO and superoxide, and the resulting generation of peroxynitrite can contribute to cytotoxicity. This mechanism has been confirmed in neuronal cultures, as well as in macrophages which express iNOS [9,10]. It is conceivable that non-enzymatic sources of NO generation may also contribute to peroxynitrite formation: such mechanisms may include the reduction of nitrite to NO under acidic conditions [11] and the non-enzymatic generation of NO from L-arginine and hydrogen peroxide [12]. The clarification of the importance of these latter non-enzymatic processes requires further investigations.

Cellular actions of peroxynitrite

(A) Primary cytotoxic mechanisms

The ratio of superoxide and NO determines the apparent reactivity of peroxynitrite: excess NO reduces the oxidation elicited by peroxynitrite [13-16]. The oxidant reactivity of peroxynitrite is mediated by an intermediate with the biological activity of hydroxyl radical, which is probably peroxynitrous acid or its activated isomer [2,3]. The cellular effects of peroxynitrite can be classified as primary and secondary effects (Table 1.)

Peroxynitrite is highly reactive, and can induce the oxidation of sulfhydryl groups and thioethers, and nitration and hydroxylation of aromatic compounds, e.g. tyrosine, tryptophan and guanine[17-20]. These and other reactions underlie the molecular basis of the cellular toxicity of peroxynitrite. Peroxynitrite has been shown to inhibit manganese superoxide dismutase, tyrosine hydroxylase, membrane Na^+/K^+ ATP-ase, membrane sodium channels glyceraldehyde-3-phosphate dehydrogenase, mitochondrial and cytosolic aconitase, as well as a number of critical enzymes in the mitochondrial respiratory chain [21-30].

Table 1. Primary and secondary effects of peroxynitrite in cells and tissues

Primary effects	Mechanism of action
Damage to lipids	Peroxidation
Glutathione depletion	Oxidation
DNA strand breakage	Oxidation
Inactivation of superoxide dismutase	Tyrosine nitration
Inhibition of respiratory enzymes	Oxidation
Activation of cyclooxygenase	

Secondary effects	Mechanism of action
Activation of PARS	DNA single strand breakage
Inhibition of cellular energetics	Inhibition of mitochondrial enzymes Inhibition of GAPDH Inhibition of Na^+/K^+ ATP-ase Depletion of glutathione PARS activation
NAD^+ and ATP depletion	Activation of PARS Inactivation of mitochondria
Disturbances in signal transduction	Interaction with critical sulfhydryl groups, Tyrosine nitration
Apoptosis	DNA strand breakage Inhibition of cell respiration

In addition to the interactions of peroxynitrite with proteins, an important interaction of peroxynitrite occurs with nucleic acids [31]. Two main types of reactions have been described: DNA base modifications and DNA single strand breakage. The reported base modifications include the formation of 8-nitroguanine, 8 oxoguanine 4,5-dihydro-5-hydroxy-4-(nitrosooxy)-2'-deoxyguanosine and oxidized and deaminated base products, such as 5-hydroxyhydantoin, 5-(hydroxymethyl)uracil, thymine glycol, 4,6-diamino-5-formamidepyrimidine, 2.6-diamino-5-formamidepyrimidine, 8-oxoadenine, 8-oxoguanine, hypoxanthine and xanthine [31-34]. The mechanism of the peroxynitrite-induced DNA single strand breakage [35-37] is probably related to abstraction of hydrogen atoms from the ribose of the DNA moiety, thereby opening the sugar ring.

(B) Activation of poly (ADP-ribose) synthetase and other secondary mechanism of cytotoxicity

Some of the cellular actions of peroxynitrite are summarized in Table 1. These actions can be classified as primary and secondary cytotoxic mechanisms. Primary mechanisms include direct oxidations and nitrations. These primary mechanisms sometimes can trigger secondary cellular responses. For example, inhibition of mitochondrial respiration can lead to "leak" of superoxide from the mitochondria, which, in turn, induces secondary cellular injury [38]. This superoxide can, in turn, initiate the formation of additional peroxynitrite. Another example of a secondary cytotoxic mechanism is the activation of the nuclear enzyme poly (ADP-ribose) synthetase, which occurs due to peroxynitrite-generated single strands in the DNA (see below). Yet another secondary mechanism of the peroxynitrite-triggered apoptotic cell death.

Activation of the nuclear enzyme poly-ADP ribosyl synthetase (PARS) leads to a massive cellular energetic deficit, and cell death via the necrotic pathway [31,39,40]. Briefly, DNA single strand breakage is the obligatory trigger of activation of PARS. When activated, PARS catalyses the cleavage of NAD^+ into ADP-ribose and nicotinamide. PARS covalently attaches ADP-ribose to various nuclear proteins, such as histones and PARS itself. Activation of PARS can rapidly deplete NAD^+, slowing the rate of glycolysis, electron transport, and ATP formation, resulting in cell dysfunction and cell death. In neurons and glial cells [41-43], cultured macrophages [37,44] and cultured rat aortic smooth muscle cells [44], profound inhibition by peroxynitrite of mitochondrial respiration has been observed, with inhibition of NADH-COQ1 reductase, succinate-cytochrome c reductase and cytochrome c oxidase activities. In macrophages, smooth muscle cells, epithelial cells and endothelial cells, peroxynitrite has also been shown to cause a marked reduction of intracellular NAD^+ and ATP levels [34,45,46]. In contrast, the suppression of ATP levels by NO is not associated with depletion of cellular NAD^+ stores [46]. Pharmacological inhibition of PARS activity has been shown to protect against cell damage in response to exogenously or endogenously produced peroxynitrite [37,43,45-48].

In addition to the initiation of the necrotic pathway, peroxynitrite (lower levels, longer time of exposure) can also lead to cell death via the apoptotic pathway [49-55]. The mechanism of peroxynitrite-induced thymocyte apoptosis involves novel protein synthesis and can be blocked by actinomycin B and cycloheximide (Virag and Szabo, unpublished observations, 1997). PARS, however, does not play a role in this latter process, since inhibition of PARS does not appear to prevent peroxynitrite-induced apoptosis [53,55].

Another secondary mechanism of peroxynitrite-induced cytotoxicity may be related to the disruption of membrane signal transduction pathways by peroxynitrite. Multiple mechanisms of such interactions have recently been characterized [56-58].

Role of peroxynitrite in shock, inflammation and reperfusion injury

Peroxynitrite-induced alterations in vascular and myocardial function

In vascular tissues and isolated hearts exposed to peroxynitrite, pathophysiological alterations occur that closely resemble the alterations seen in various forms of shock, inflammation, and reperfusion injury. The peroxynitrite-induced vascular and myocardial effects can be characterized as acute or delayed changes.

The acute cardiovascular effects of peroxynitrite are rapid relaxations and inhibition of contractions [14,59]. These acute effects of peroxynitrite are likely to be related to NO, released from the NO donors formed by the reaction of glucose and peroxynitrite [60,61]. In addition, peroxynitrite may also have acute vasodilatory actions via activation of ATP-sensitive potassium channels [62].

The delayed effects are probably more relevant from a pathophysiological point of view. Peroxynitrite infusion causes a reduction in myocardial contractility in isolated perfused hearts [63,64] and induces an impairment of the endothelium-dependent relaxant ability [14]. Similarly, peroxynitrite causes an impairment of the endothelium-dependent relaxations in isolated blood vessels [55]. The peroxynitrite-mediated depression in cardiac efficiency may be due to reduced coupling between ATP production and mechanical work [64], and/or due to disturbances in intracellular calcium handling [65,66]. The delayed vascular changes in response to peroxynitrite are, at least in part, related to the PARS-mediated futile energy depleting cycle [47].

Circulatory shock

The cardiovascular consequences of circulatory shock include an NO-mediated. reduced responsiveness of arteries and veins to exogenous or endogenous vasoconstrictor agents (vascular hyporeactivity), myocardial dysfunction and disruption of intracellular energetic processes [67,68]. Recent data, demonstrating (1) the generation of peroxynitrite in various forms of shock; and (2) the capability of authentic peroxynitrite to mimic many of the cardiovascular alterations associated with shock (endothelial dysfunction, vascular hyporeactivity. myocardial failure and cellular energetic failure), raised the possibility, that part of the NO-mediated cardiovascular and metabolic changes are related to peroxynitrite. Studies with the cell-permeable superoxide dismutase analog [69] and peroxynitrite scavenger [48]compound MnIII tetrakis (4-benzoic acid) porphyrin, which demonstrated a full prevention of the depression of mitochondrial respiration in macrophages obtained from endotoxemic rats and a partial protection against the endotoxin-induced reduction of the *ex vivo* contractility of the thoracic aorta [70]. further supported the view that peroxynitrite is involved in these alterations.

The evidence that peroxynitrite may contribute to the development of vascular hypocontractility in shock prompted us to investigate the role of PARS in the vascular hyporeactivity and cellular energetic alterations associated with circulatory shock. Somewhat surprisingly, we have observed that pharmacological inhibition of PARS with 3-aminobenzamide completely prevented the loss of contractility in thoracic aortic rings obtained from rats subjected to endotoxic shock [47]. This finding may be surprising for at least two reasons: (1) the proposition that PARS activation and the resultant energetic depletion are responsible for the contractile deficit during shock contradicts a previous theory that the massive vasodilatation and vascular hyporeactivity in blood vessels that express iNOS is mediated by the NO - cGMP pathway, and (2) the protection by the PARS inhibitor was complete against the loss of contractility, which is in contrast to most of the *in vitro* data with authentic peroxynitrite, where the protection by PARS inhibitors was substantial, but incomplete.

According to *in vitro* experiments, peroxynitrite is directly cytotoxic to cultured endothelial cells, and PARS activation plays a role in mediating this cytotoxicity. For example, exposure of cultured human umbilical vein endothelial cells to authentic peroxynitrite results in reduced mitochondrial respiration in these cells, and this is reduced by pretreatment with the PARS inhibitor 3-aminobenzamide [55]. Accordingly, rat studies were designed, in which the animals were subjected to high doses of endotoxin in the absence or presence of 3-aminobenzamide. The results demonstrated that the inhibitor of PARS ameliorated the development of the endothelial dysfunction *ex vivo* [55]. Although it is clear that oxidants other than peroxynitrite may also contribute to the impaired endothelial function in endotoxic shock (such as hydroxyl radical, generated from hydrogen peroxide), the current observations, nevertheless, provide pharmacological evidence for the role of PARS activation in the development of endothelial dysfunction in endotoxic shock.

Inflammation

Carrageenan and zymosan induced paw edema and pleurisy models are convenient and rapid methods for the investigation of inflammatory responses. The cellular and molecular mechanism of these inflammatory responses are well characterized, and these models of inflammation are standard models of screening for anti-inflammatory activity of various experimental compounds. The early phase of the inflammation is related to the production of histamine, leukotrienes, platelet-activating factor, and possibly cyclooxygenase products, while the delayed phase of the inflammatory response has been linked to neutrophil infiltration and the production of neutrophil-derived free radicals, such as hydrogen peroxide, superoxide and hydroxyl radical, as well as to the release of other neutrophil-derived mediators [71-73]. Recent studies have demonstrated the production of peroxynitrite in these local models of inflammation [73,74]. In addition, using NOS inhibitors, superoxide dismutase mimetics and other antioxidants, a connection between peroxynitrite generation and the inflammatory response has been put forward [73,74]. Similarly to the *in vitro* studies, using pharmacological inhibitors of PARS and PARS knockout animals, we have demonstrated that DNA single strand breakage and poly (ADP-ribose) synthetase activation, which contribute to the development of these inflammatory responses [74,75].

Importantly, inhibition of PARS during inflammation was associated with reduced neutrophil recruitment into the inflamed organs [74,75]. The mechanism of reduced neutrophil issue infiltration in the presence of PARS inhibitors is unclear, but may be related to (1) modulation of a post-adhesion phenomenon (as demonstrated by intravital microscopy studies using the PARS inhibitor 3-aminobenzamide) [74], (2) prevention of the impairment of the endothelial function during inflammation (Szabó et al., 1997) and/or (3) regulation of the expression of adhesion molecules by PARS [76,77]. By suppressing the infiltration of neutrophils into the inflamed or reperfused tissues, inhibitors of PARS may exert important anti-inflammatory actions (Fig. 1).

The formation of peroxynitrite has been demonstrated in a number of inflammatory conditions, such as arthritis [78,79], ileitis [80], inflammatory bowel disease [81], uveitis [82], endotoxin-induced intestinal inflammation [83], allergic encephalomyelitis [84] and autoimmune diabetes [85]. Increased nitrotyrosine staining was also found in inflamed human brain [86,87], myocardial [88], and intestinal [89] tissues. The ability of authentic peroxynitrite to cause severe inflammation has also been confirmed [90].

The elucidation of the role of PARS activation in the above inflammatory conditions requires further investigations. Preliminary data, nevertheless, indicate that pharmacological inhibition of PARS protects against the course of disease progression in animal models of arthritis [91,92]; allergic encephalomyelitis [93]; and inflammatory bowel disease [94]. There are also a number of convincing *in vitro* data implicating the role of PARS activation in the pathophysiology of inflammatory and oxidant-mediated pancreatic islet cell destruction and diabetes development [95-99]. However, it is not clear, at present, whether peroxynitrite, or other oxidant mediators of inflammation (such as hydroxyl radical) are responsible for the DNA single strand breakage, which precedes the activation of PARS, in the above mentioned pathophysiological states.

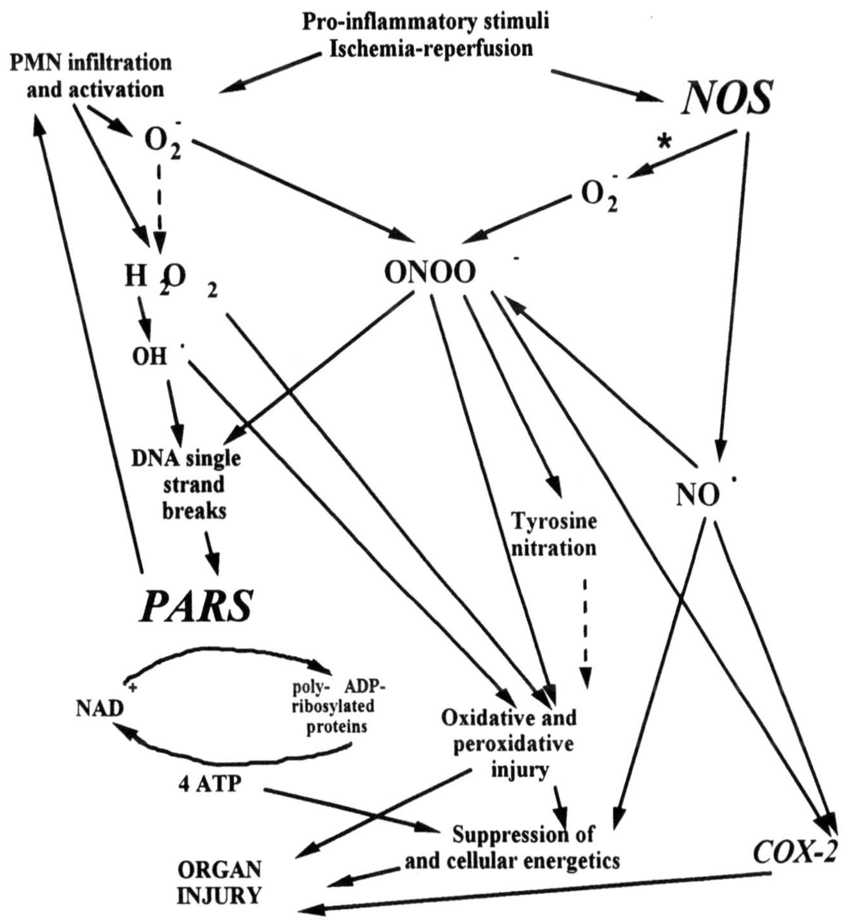

Fig. 1. Delayed pathways of cell injury involving nitric oxide (NO$^\cdot$), superoxide ($O_2^{\cdot-}$), hydroxyl radical (OH$^\cdot$) and peroxynitrite (ONOO$^-$). The source of NO can be constitutive NO synthesis, but at later stages, expression of iNOS can occur. Under conditions of low cellular L-arginine (*), NOS may produce both superoxide *and* NO. NO combines with superoxide to yield peroxynitrite. Part of the injury by hydroxyl radical and peroxynitrite is related to the development of DNA single strand breakage, with consequent activation of PARS. PARS activation is involved in the facilitation of neutrophils into the inflammatory site. Peroxynitrite or NO can increase the catalytic activity of cyclooxygenase (COX-2).

Ischemia-reperfusion injury

Anoxia-reoxygenation of cultured cells induces the generation of peroxynitrite [100, 101]. Similarly, in ischemia-reperfusion injury, superoxide, produced in the reperfusion phase, has reacts with NO to form peroxynitrite. This has been demonstrated in many reperfused organs, including the heart, liver, intestine, brain and lung [102-106]. In these conditions, prevention of peroxynitrite generation by inhibition of NO biosynthesis markedly reduces reperfusion injury [63,103,104].

A growing body of evidence supports the role of the peroxynitrite in the cellular injury associated with ischemia-reperfusion injury. For instance, in the central nervous system, peroxynitrite (and not NO or superoxide, independently), appears to be the major cytotoxic mediator in the neuronal injury during stroke and N-methyl-D-aspartate (NMDA) receptor activation [106-109]. In the intestine, reperfusion injury begins with the accumulation of hypoxanthine from ATP metabolism and the conversion of xanthine dehydrogenase to xanthine oxidase. The latter enzyme catalyzes the conversion of hypoxanthine with concomitant generation of superoxide radicals [110]. Because peroxynitrite enhances the conversion of xanthine oxidase from xanthine dehydrogenase [111], a positive feedback cycle between peroxynitrite and xanthine dehydrogenase/xanthine oxidase may be present in reperfusion injury. Studies in rats have demonstrated that reperfusion of the ischemic intestine is associated with generation of peroxynitrite [15,105]. Peroxynitrite is capable of increasing epithelial permeability, a process that can be blocked by PARS inhibitors [45,46]. Accordingly, *in vivo* inhibition of PARS during reperfusion ameliorates reperfusion-induced intestinal hyperpermeability[112]. Moreover, during reperfusion of the ischemic intestine, inhibition of PARS reduces neutrophil recruitment (in line with the results of the studies in inflammation, see above), and improves survival rate [105].

We have recently demonstrated that the peroxynitrite-PARS pathway also plays an important role in the pathogenesis of myocardial reperfusion injury: pharmacological inhibition of PARS reduces infarct size, and attenuates neutrophil infiltration during reperfusion [113] Inhibition of PARS also attenuates nitrotyrosine staining in the reperfused myocardium [113]. This latter effect is not due to a direct scavenging effect of the PARS inhibitor used, but, rather, related to interruption of a positive feedback cycle involving neutrophils and peroxynitrite generation [113]. A similar positive feedback cycle is also present in zymosan and carrageenan models of inflammation, and in splanchnic reperfusion injury [74,75,105].

The mechanism of peroxynitrite-induced reperfusion injury during stroke appear to involve multiple mechanisms, but recent studies have established the crucial importance of the PARS pathway [43,114-117]. Notably, recent studies by Snyder's and Moskowitz's groups have demonstrated that genetic ablation of PARS renders animals resistant to cerebral infarction following middle cerebral artery occlusion and reperfusion injury [43,117]. The degree of protection in these studies is remarkable: in the PARS knockout animals, the infarct size was approximately 20% of the one developed in wild-type control animals.

Conclusions - future directions

Neither NO, nor superoxide alone, act as strong oxidants towards most types of organic compounds. The evidence presented above favors the view that the reaction of NO and superoxide yields peroxynitrite, which, under many conditions, enhances the cytotoxic potential of its "precursors". Although chemical considerations favor the production of peroxynitrite, one has to emphasize that the actual demonstration of the presence or production of peroxynitrite in pathophysiological conditions is far from straightforward. Peroxynitrite rapidly oxidizes the fluorescent probe dihydrorhodamine 123 to rhodamine 123 *in vitro* [118]. The production of peroxynitrite can be evidenced as increased oxidation of dihydrorhodamine 123 to rhodamine 123 in the plasma [15]. Caution should be exercised with this method, for oxidation of dihydrorhodamine can be triggered by oxidants other than peroxynitrite (hydroxyl radical, for example). A NOS-inhibitor inhibitable component of the oxidation of dihydrorhodamine can be taken as evidence of an effect of peroxynitrite [15,118]. Nitrotyrosine formation, and its detection by immunostaining, was initially proposed as a relatively specific means for detection of the "footprint" of peroxynitrite [18]. Recent evidence indicates, however, that certain other reactions can also induce tyrosine nitration [119-121]: increased nitrotyrosine staining as an indication of "increased nitrosative stress", rather than a specific marker of peroxynitrite. Specific peroxynitrite scavengers which could help delineating the role of peroxynitrite in circulatory shock, or in any other pathophysiological condition are not available. Uric acid, a putative scavenger of peroxynitrite, [10,47,100,118,122,123], can scavenge other oxidants as well [124]. Therefore, the evidence implicating the role of peroxynitrite in a given pathophysiological condition can only be indirect. A simultaneous protective effect of superoxide neutralizing strategies and NO synthesis inhibition, coupled with the demonstration of peroxynitrite formation in the particular pathophysiological condition, can be taken as strong indication for the role of peroxynitrite.

It appears that a major novel mechanism of peroxynitrite-induced cytotoxicity is related to DNA single strand breakage and PARS activation. PARS inhibition is clearly of therapeutic benefit in a variety of pathophysiological conditions. PARS knockout animals are alive and viable, and do not show signs of disease [125,126], although they may be more susceptible to radiation-induced injury [126]. Therefore, long-term inhibition of PARS, using appropriate pharmacological tools, probably will not induce major side effects. Taken together, based on the recent breakthroughs on the final cellular effectors of free radical- and oxidant-mediated injury, the following novel therapeutic approaches (alone, or in combination) can be proposed, for the experimental therapy of shock, inflammation and reperfusion injury: (1) scavenging of peroxynitrite, (2) inhibition of the production or action of superoxide and other oxygen-derived free radicals (e.g. with cell-permeable superoxide dismutase analogs), (3) selective inhibition of iNOS and (4) pharmacological inhibition of PARS.

Acknowledgement: This work was supported, in part by a Grant-In-Aid from the American Heart Association (#96006400), a Grant-In-Aid from the American Lung Association (RG033N) and by a Grant from the National Institutes of Health (R29GM54773).

References

1. Beckman JS, Beckman TW, Chen J, Marshall PA, Freeman BA (1990) Apparent hydroxyl radical production by peroxynitrite: implication for endothelial injury from nitric oxide and superoxide. Proc Natl Acad Sci USA 87: 1620-1624
2. Pryor W, Squadrito G (1995) The chemistry of peroxynitrite: a product from the reaction of nitric oxide with superoxide. Am J Physiol 268: L699-L722
3. Beckman JS, Koppenol WH (1996) Nitric oxide, superoxide, and peroxynitrite: the good, the bad, and ugly. Am J Physiol 271: C1424-37
4. McCord J (1993) Oxygen-derived free radicals. New Horizons 1: 70-76
5. Nohl H (1994) Generation of superoxide radicals as byproduct of cellular respiration. Ann Biol Clin 52: 199-204
6. Chan PH (1996) Role of oxidants in ischemic brain damage. Stroke 27: 1124-9, 1996
7. Halliwell B (1996) Mechanisms involved in the generation of free radicals. Pathol Biol 44: 6-13
8. Szabó C (1996) The role of peroxynitrite in the pathophysiology of shock, inflammation and ischemia-reperfusion injury. Shock 6: 79-88
9. Xia Y, Dawson VL, Dawson TM, Snyder SH, Zweier JL (1996) Nitric oxide synthase generates superoxide and nitric oxide in arginine-depleted cells leading to peroxynitrite-mediated cellular injury. Proc Natl Acad Sci USA 93: 6770-4
10. Xia Y, Zweier JL (1997) Superoxide and peroxynitrite generation from inducible nitric oxide synthase in macrophages. Proc Natl Acad Sci USA 94: 6954-8
11. Zweier JL, Wang P, Samouilov A, Kuppusamy P (1995) Enzyme-independent formation of nitric oxide in biological tissues. Nature Med 1: 804-809

12. Nagase S, Takemura K, Ueda A., Hirayama A, Aoyagi K, Kondoh M, Koyama A (1997) A novel nonenzymatic pathway for the generation of nitric oxide by the reaction of hydrogen peroxide and D- or L-arginine. Biochem Biophys Res Comm 233: 150-153
13. Rubbo H, Radi R, Trujillo M, Telleri R, Kalyanaraman B, Barnes S, Kirk M, Freeman BA (1994) Nitric oxide regulation of superoxide and peroxynitrite-dependent lipid peroxidation. Formation of novel nitrogen-containing oxidized lipid derivatives. J Biol Chem 269: 26066-75
14. Villa LM, Salas E, Darley-Usmar M, Radomski MW, Moncada S (1994) Peroxynitrite induces both vasodilatation and impaired vascular relaxation in the isolated perfused rat heart. Proc Natl Acad Sci USA 91:12383-12387
15. Szabó C, Salzman AL, Ischiropoulos H (1995) Peroxynitrite-mediated oxidation of dihydrorhodamine 123 occurs in early stages of endotoxic and hemorrhagic shock and ischemia-reperfusion injury. FEBS Lett 372:229-232
16. Miles AM, Bohle DS, Glassbrenner PA, Hansert B, Wink DA, Grisham MB (1996) Modulation of superoxide-dependent oxidation and hydroxylation reactions by nitric oxide. J Biol Chem 271:40-47
17. Radi R, Beckman JS, Bush KM, Freeman BA (1991) Peroxynitrite-induced membrane lipid peroxidation: the cytotoxic potential of superoxide and nitric oxide. Arch Biochem Biophys 288: 481-7
18. Ischiropoulos H, Zhu L, Chen J, Tsai M, Martin JC, Smith CD, Beckman JS (1992) Peroxynitrite-mediated tyrosine nitration catalyzed by superoxide dismutase. Arch Biochem Biophys 298:431-7
19. Alvarez B, Rubbo H, Kirk M, Barnes S, Freeman BA, Radi R (1996) Peroxynitrite-dependent tryptophan nitration. Chem Res Toxicol 9: 390-6
20. Beckman JS (1996) Oxidative damage and tyrosine nitration from peroxynitrite. Chem Res Toxicol 9: 836-44
21. Bauer ML, Beckman JS, Bridges RJ, Fuller CM, Matalon S (1992) Peroxynitrite inhibits sodium uptake in rat colonic membrane vesicles. Biochim Biophys Acta 1104:87-94
22. Ischiropoulos H, Duran D, Horwitz J (1995) Peroxynitrite-mediated inhibition of DOPA synthesis in PC12 cells. J Neurochem 65:2366-2372
23. Castro L, Rodriguez M, Radi R (1994) Aconitase is readily inactivated by peroxynitrite, but not by its precursor, nitric oxide. J Biol Chem 269:29409-29415
24. Mohr S, Stamler JS, Brune B (1994) Mechanism of covalent modification of glyceraldehyde-3-phosphate dehydrogenase at its active site thiol by nitric oxide, peroxynitrite and related nitrosating agents. FEBS Lett. 348:223-7
25. Hu P, Ischiropoulos H, Beckman JS, Matalon S (1994) Peroxynitrite inhibition of oxygen consumption and sodium transport in alveolar type II cells. Am J Physiol 266:L628-634
26. Radi R, Rodriguez M, Castro L, Telleri R (1994) Inhibition of mitochondrial electron transport by peroxynitrite. Arch Biochem Biophys 308: 89-95
27. Crow JP, Beckman JS, McCord JM (1995) Sensitivity of the essential zinc-thiolate moiety of yeast alcòhol dehydrogenase to hypochlorite and peroxynitrite. Biochemistry 34: 3544-52
28. Cassina A, Radi R (1996) Differential inhibitory action of nitric oxide and peroxynitrite on mitochondrial electron transport. Arch Biochem Biophys 328: 309-16

29. Sato T, Kamata Y, Irifune M, Nishikawa T (1997) Inhibitory effect of several nitric oxide-generating compounds on purified Na+,K(+)-ATPase activity from porcine cerebral cortex. J Neurochem 68: 1312-8
30. MacMillan-Crow LA, Crow JP, Kerby JD, Beckman JS, Thompson JA (1996) Nitration and inactivation of manganese superoxide dismutase in chronic rejection of human renal allografts. Proc Natl Acad Sci USA 93: 11853-8
31. Szabó C, Ohshima H (1998) DNA injury induced by peroxynitrite. Nitric Oxide Biol Chem, in press
32. Inoue S, Kawanishi S (1995) Oxidative DNA damage induced by simultaneous generation of nitric oxide and superoxide. FEBS Lett 371:86-88
33. Yermilow V, Rubio J, Becchi M, Friesen MD, Pignatelli B, Ohshima H (1995) Formation of 8-nitroguanine by the reaction of guanine with peroxynitrite *in vitro*. Carcinogenesis 16: 2045-2050
34. Yermilov V, Yoshie Y, Rubio J, Ohshima H (1996) Effects of carbon dioxide/bicarbonate on induction of DNA single-strand breaks and formation of 8-nitroguanine, 8-oxoguanine and base-propenal mediated by peroxynitrite. FEBS Lett 399: 67-70
35. King PA, Anderson VE, Edwards JO, Gustav G, Plumb RC, Suggs JW (1992) A stable solid that generates hydroxyl radical dissolutions in aqueous solutions: reaction with proteins and nucleic acid. J Am Chem Soc 114: 5430-5432
36. Salgo MG, Bermudez E, Squadrito G, Pryor W (1995) DNA damage and oxidation of thiols peroxynitrite causes in rat thymocytes. Arch Biochem Biophys 322: 500-505
37. Szabó C, Zingarelli B, O'Connor M, Salzman, A.L (1996) DNA strand breakage, activation of poly-ADP ribosyl synthetase, and cellular energy depletion are involved in the cytotoxicity in macrophages and smooth muscle cells exposed to peroxynitrite. Proc Natl Acad Sci USA 93: 1753-1758
38. Packer MA, Porteous CM, Murphy MP (1996) Superoxide production by mitochondria in the presence of nitric oxide forms peroxynitrite. Biochem Mol Biol Int 40: 527-34
39. Cochrane CG (1991) Mechanisms of oxidant injury of cells. Molec Aspects Med 12:137-147
40. Szabó C (1996) DNA strand breakage and activation of poly-ADP ribosyltransferase: a cytotoxic pathway triggered by peroxynitrite. Free Radical Biol Med 21: 855-869
41. Bolanos JP, Heales SJ, Land JM, Clark JB (1995) Effect of peroxynitrite on the mitochondrial respiratory chain: differential susceptibility of neurones and astrocytes in primary culture. J Neurochem 64:1965-72
42. Zhang J, Dawson VL, Dawson TM, Snyder SH (1994) Nitric oxide activation of poly (ADP-ribose) synthetase in neurotoxicity. Science 263: 687-689
43. Eliasson MJL, Sampei K, Mandir AS, Hurn PD, Traystman RJ, Bao J, Pieper A, Wang ZQ, Sdawson TM, Snyder SH, Dawson VL (1997) Poly (ADP-ribose) polymerase gene disruption renders mice resistant to cerebral ischemia. Nature Med 3: 1089-1095
44. Zingarelli B, O'Connor M, Wong H, Salzman AL, Szabó C (1996) Peroxynitrite-mediated DNA strand breakage activates poly-ADP ribosyl synthetase and causes cellular energy depletion in macrophages stimulated with bacterial lipopolysaccharide. J Immunol 156: 350-358
45. Szabó C, Saunders C, O'Connor M, Salzman AL (1997) Peroxynitrite causes energy depletion and increases permeability via activation of poly-ADP ribosyl synthetase in pulmonary epithelial cells. Am J Resp Mol Cell Biol 16: 105-109

46. Kennedy M, Szabó C, Salzman AL (1998) Activation of polyADP ribosyl synthetase (PARS) mediates cytotoxicity induced by peroxynitrite in human intestinal epithelial cells. Gastroenterology, in press
47. Szabó C , Zingarelli B, Salzman AL (1996) Role of poly-ADP ribosyltransferase activation in the nitric oxide- and peroxynitrite-induced vascular failure. Circ Res 78: 1051-1063
48. Szabó C, Day BJ, Salzman AL (1996) Evaluation of the relative contribution of nitric oxide and peroxynitrite to the suppression of mitochondrial respiration in immunostimulated macrophages, using a novel mesoporphyrin superoxide dismutase analog and peroxynitrite scavenger. FEBS Lett 381: 82-86.
49. Bonfoco E, Krainc D, Ankarcrona M, Nicotera P, Lipton S (1995) Apoptosis and necrosis: two distinct events induced, respectively, by mild and intense insults with NMDA or nitric oxide/superoxide in cortical cell cultures. Proc Natl Acad Sci USA 92: 7162-7166
50. Estevez AG, Radi R, Barbeito L, Shin JT, Thompson JA, Beckman JS (1995) Peroxynitrite-induced cytotoxicity in PC12 cells: evidence for an apoptotic mechanism differentially modulated by neurotropic factors. J Neurochem 65: 1543-1550
51. Salgo MG, Squadrito GL, Pryor WA (1995) Peroxynitrite causes apoptosis in rat thymocytes. Biochem Biophys Res Comm 215: 1111-1118
52. Sandoval M, Zhang XJ, Liu X, Mannick EE, Clark DA, Miller MJ (1996) Peroxynitrite-induced apoptosis in T84 and RAW 264.7 cells: attenuation by L-ascorbic acid. Free Rad Biol Med 22: 489-95
53. Leist M, Single B, Kunstle G, Volbracht C, Hentze H, Nicotera P (1997) Apoptosis in the absence of poly-(ADP-ribose) polymerase. Biochem Biophys Res Comm 233: 518-522
54. O'Connor M, Salzman AL, Szabó C (1997) Role of peroxynitrite in the protein oxidation and apoptotic DNA fragmentation in immunostimulated vascular smooth muscle cells. Shock, in press
55. Szabó C, Cuzzocrea S, Zingarelli B, O'Connor M, Salzman AL (1997) Endothelial dysfunction in endotoxic shock: importance of the activation of poly (ADP ribose) synthetase (PARS) by peroxynitrite. J Clin Invest 100: 723-735
56. Gow AJ, Duran D, Malcolm S, Ischiropoulos H (1996) Effects of peroxynitrite-induced protein modifications on tyrosine phosphorylation and degradation. FEBS Lett 385: 63-6
57. Berlett BS, Friguet B, Yim MB. Chock PB, Stadtman ER (1996) Peroxynitrite-mediated nitration of tyrosine residues in Escherichia coli glutamine synthetase mimics adenylylation: relevance to signal transduction. Proc Natl Acad Sci USA 93: 1776-80, 1996
58. Kong SK, Yim MB, Stadtman ER, Chock PB (1996) Peroxynitrite disables the tyrosine phosphorylation regulatory mechanism: Lymphocyte-specific tyrosine kinase fails to phosphorylate nitrated cdc2(6-20)NH2 peptide. Proc Natl Acad Sci USA 93: 3377-82
59. Wu M, Pritchard KA Jr, Kaminski PM, Fayngersh RP, Hintze TH, Wolin MS (1994) Involvement of nitric oxide and nitrosothiols in relaxation of pulmonary arteries to peroxynitrite. Am J Physiol 266: H2108-13
60. Tarpey MM, Beckman JS, Ischiropoulos H, Gore JZ, Brock TA (1995) Peroxynitrite stimulates vascular smooth muscle cell cyclic GMP synthesis. FEBS Lett. 364: 314-8

61. Mayer B, Schrammel A, Klatt P, Koesling D, Schmidt K (1995) Peroxynitrite-induced accumulation of cyclic GMP in endothelial cells and stimulation of purified soluble guanylyl cyclase. Dependence on glutathione and possible role of S-nitrosation. J Biol Chem 270:17355-60
62. Wei EP, Kontos HA, Beckman JS (1996) Mechanisms of cerebral vasodilation by superoxide, hydrogen peroxide, and peroxynitrite. Am J Physiol 271: H1262-6
63. Yasmin W, Strynadka KD, Schulz R (1997) Generation of peroxynitrite contributes to ischemia-reperfusion injury in isolated rat hearts. Cardiovasc Res 33: 422-32
64. Schulz R, Dodge KL, Lopaschuk GD, Clanachan AS (1997) Peroxynitrite impairs cardiac contractile function by decreasing cardiac efficiency. Am J Physiol 272: H1212-9, 1997
65. Ishida H, Ichimori K, Hirota Y, Fukahori M, Nakazawa H (1996) Peroxynitrite-induced cardiac myocyte injury. Free Rad Biol Med 20: 343-50, 1996
66. Viner RI, Huhmer AF, Bigelow DJ, Schoneich C (1996) The oxidative inactivation of sarcoplasmic reticulum Ca(2+)-ATPase by peroxynitrite. Free Radical Res 24: 243-59
67. Szabó C, Thiemermann C (1994) Role of nitric oxide in haemorrhagic, traumatic and anaphylactic shock and in thermal injury. Shock 2: 145-155
68. Kilbourn RG, Traber D, Szabó C (1997) Role of nitric oxide in the vascular failure in shock. Dis Month 43: 277-348
69. Faulkner KM, Liochev SI, Fridowich I (1994) Stable Mn(III) prophyrins mimic cuperoxide dismutase *in vitro* and substitute for it *in vivo*. J Biol Chem 269:23471-23476
70. Zingarelli B, Day BJ, Crapo J, Salzman AL, Szabó C (1997) The potential involvement of peroxynitrite in the pathogenesis of endotoxic shock. Br J Pharmacol 120: 259-267
71. DiRosa M, Willoughby DA (1971) Screens for anti-inflammatory drugs. J Pharm Pharmacol 23: 297-300
72. Dawson J, Sedgwick AD, Edwards JC, Lees P (1991) A comparative study of the cellular, exudative and histological responses to carrageenan, dextran and zymosan in the mouse. Int J Tissue React 13: 171-85
73. Salvemini D, Wang ZQ, Wyatt DM, Bourdon MH, Marino PT, Currie MG (1996) Nitric oxide: a key mediator in the early and late phase of carrageenan-induced rat paw inflammation. Br J Pharmacol 118: 829-838
74. Szabó C, Lim LH, Cuzzocrea S, Getting SJ, Zingarelli B, Flower RJ, Salzman AL, Perretti M (1997) Inhibition of poly (ADP-ribose) synthetase exerts anti-inflammatory effects and inhibits neutrophil recruitment. J Exp Med 186: 1041-1049
75. Cuzzocrea S, Zingarelli B, Gilad E, Hake P, Salzman AL, Szabó C (1998) Protective effects of 3-aminobenzamide, an inhibitor of poly (ADP-ribose) synthase in a carrageenan-induced model of local inflammation. Eur J Pharmacol, in press
76. Roebuck KA, Rahman A, Lakshminarayanan V, Janakidevi K, Malik AB (1995) H2O2 and tumor necrosis factor-alpha activate intercellular adhesion molecule 1 (ICAM-1) gene transcription through distinct cis-regulatory elements within the ICAM-1 promoter. J Biol Chem 270: 18966-74
77. Szabó C, Zingarelli B, Cuzzocrea S, Salzman AL (1997) Poly (ADP-ribose) synthetase modulates expression of P-selectin and ICAM-1 in myocardial ischemia-reperfusion injury. Jp J Pharmacol 75: (Suppl) 101P
78. Kaur H, Halliwell B (1994) Evidence for nitric oxide-mediated oxidative damage in chronic inflammation, Nitrotyrosine in serum and synovial fluid from rheumatoid patients. FEBS Lett 350: 9-12

79. Hukkanen M, Corbett SA, Batten J, Konttinen YT, McCarthy ID, Maclouf J, Santavirta S, Hughes SP, Polak JM (1997) Aseptic loosening of total hip replacement. Macrophage expression of inducible nitric oxide synthase and cyclo-oxygenase-2, together with peroxynitrite formation, as a possible mechanism for early prosthesis failure. J Bone Joint Surg Br 79: 467-74
80. Miller MJS, Thompson JH, Zhang XJ, Sadowska-Krowicka H, Kakkis JL, Munshi UK, Sandoval M, Rossi J, Elobi-Childress S, Beckman J, Ye YZ, Rodi CP, Manning P, Currie M, Clark DA (1995) Role of inducible nitric oxide synthase expression and peroxynitrite formation in guinea pig ileitis. Gastroenterology 109:1475-1483
81. Singer II, Kawka DW, Scott S, Weidner JR, Mumford RA, Riehl TE, Stenson WF (1996) Expression of inducible nitric oxide synthase and nitrotyrosine in colonic epithelium in inflammatory bowel disease. Gastroenterology. 111: 871-85
82. Wu GS, Zhang J, Rao NA (1997) Peroxynitrite and oxidative damage in experimental autoimmune uveitis. Invest Ophthalmol Vis Sci 38: 1333-9
83. Chamulitrat W, Skrepnik NV, Spitzer JJ (1997) Endotoxin-induced oxidative stress in the rat small intestine: role of nitric oxide. Shock 5: 217-22, 1996
84. Van Der Veen R, Hinto DR, Incardona F, Hofman FM (1997) Extensive peroxynitrite activity during progressive stages of central nervous system inflammation. J Neuroimmunol 77: 1-7
85. Suarez-Pinzon WL, Szabó C, Rabinovitch A (1997) Development of autoimmune diabetes in NOD mice is associated with the formation of peroxynitrite in pancreatic islet beta-cells. Diabetes 46: 907-911
86. Bagasra O, Michaels FH, Zheng YM, Bobroski LE, Spitsin SV, Fu ZF, Tawadros R, Koprowski H (1995) Activation of the inducible form of nitric oxide synthase in the brains of patients with multiple sclerosis. Proc Natl Acad Sci USA 92: 12041-5
87. Smith MA, Richey Harris PL, Sayre LM, Beckman JS, Perry G (1997) Widespread peroxynitrite-mediated damage in Alzheimer's disease. J Neurosci 17: 2653-7
88. Kooy NW, Lewis SJ, Royall JA, Ye YZ, Kelly DR, Beckman JS (1997) Extensive tyrosine nitration in human myocardial inflammation: evidence for the presence of peroxynitrite. Critical Care Med 25: 812-9
89. Ford H, Watkins S, Reblock K, Rowe M (1997) The role of inflammatory cytokines and nitric oxide in the pathogenesis of necrotizing enterocolitis. J Pediatric Surg 32: 275-82
90. Rachmilewitz D, Stamler JS, Karmeli F, Mullins ME, Singel DJ, Loscalzo J, Xavier RJ, Podolsky DK (1993) Peroxynitrite-induced rat colitis--a new model of colonic inflammation. Gastroenterology 105: 1681-8
91. Miesel R, Kurpisz M, Kroger H (1995) Modulation of inflammatory arthritis by inhibition of poly(ADP ribose) polymerase. Inflammation. 19: 379-87
92. Szabó C, Cuzzocrea S, Hake P, Scott GS, Hirsch R, Salzman AL (1997) Protective effect of an inhibitor of poly (ADP-ribose) synthetase in collagen-induced arthritis. Jp J Pharmacol 75: (Suppl) 102P.
93. Scott, GS, Hake P, Salzman AL, Szabó C (1998) Suppression of experimental allergic encephalomyelitis by administration of 5-iodo-6-amino-1,2,benzopyrone, a novel inhibitor of poly (ADP-ribose) synthetase. Crit Care Med, in press
94. Salzman AL, Zingarelli B, Cuzzocrea S, Szabó C. (1997) Role of peroxynitrite and poly (ADP-ribose) synthetase activation in experimental colitis. Jp J Pharmacol 75 (Suppl): 15P.

95. Yamamoto H, Uchigata Y, Okamoto H (1981) Streptozotocin and alloxan induce DNA strand breaks and poly(ADP-ribose) synthetase in pancreatic islets. Nature 294: 284-6
96. Uchigata Y, Yamamoto H, Kawamura A, Okamoto H (1982) Protection by superoxide dismutase, catalase, and poly(ADP-ribose) synthetase inhibitors against alloxan- and streptozotocin-induced islet DNA strand breaks and against the inhibition of proinsulin synthesis. J Biol Chem 257: 6084-8
97. Masiello P, Cubeddu TL, Frosina G, Bergamini E (1985) Protective effect of 3-aminobenzamide, an inhibitor of poly (ADP-ribose) synthetase, against streptozotocin-induced diabetes. Diabetologia 28: 683-6
98. Radons J, Heller B, Burkle A, Hartmann B, Rodriguez ML, Kroncke KD, Burkart V, Kolb H (1994) Nitric oxide toxicity in islet cells involves poly (ADP-ribose) polymerase activation and concomitant NAD depletion. Biochem Biophys Res Comm 199: 1270-1277
99. Heller B. Wang ZQ, Wagner EF, Radons J, Burkle A, Fehsel K, Burkart V, Kolb H (1995) Inactivation of the poly(ADP-ribose) polymerase gene affects oxygen radical and nitric oxide toxicity in islet cells. J Biol Chem 270: 11176-80
100. Xie YW, Wolin MS (1996) Role of nitric oxide and its interaction with superoxide in the suppression of cardiac muscle mitochondrial respiration. Involvement in response to hypoxia/reoxygenation. Circulation. 94: 2580-6
101. Zulueta JJ, Sawhney R, Yu FS, Cote CC, Hassoun PM (1997) Intracellular generation of reactive oxygen species in endothelial cells exposed to anoxia-reoxygenation. Am J Physiol 272: L897-902
102. Matheis G, Sherman MP, Buckberg GD, Haybron DM, Young HN, Ignarro LJ (1992) Role of L-arginine - nitric oxide pathway in myocardial reoxygenation injury. Am J Physiol 262:H616-620
103. Schulz R, Wambolt R (1995) Inhibition of nitric oxide synthesis protects the isolated working rabbit heart from ischaemia-reperfusion injury. Cardiovasc Res 30:432-439
104. Ma TT, Ischiropoulos H, Brass CA (1995) Endotoxin-stimulated nitric oxide production increases injury and reduces rat liver chemiluminescence during reperfusion. Gastroenterology 108: 463-9
105. Cuzzocrea S, Zingarelli B, Costantino G, Szabó A, Salzman AL, Caputi AP, Szabó C (1997) Beneficial effects of 3-aminobenzamide, an inhibitor of poly (ADP-ribose) synthetase in a rat model of splanchnic artery occlusion and reperfusion. Br J Pharmacol 121: 1065-1074.
106. Beckman JS (1991) The double-edged role of nitric oxide in brain function and superoxide-mediated injury. J Develop Physiol 15: 53-9
107. Dawson VL (1995) Nitric oxide: role in neurotoxicity. Clin Exp Pharmacol Physiol 22: 305-8
108. Szabó C (1996) Role of nitric oxide in the central nervous system. Brain Research Bull 41: 131-141
109. Bolanos JP, Almeida A, Stewart V, Peuchen S, Land JM, Clark JB (1997) Nitric oxide-mediated mitochondrial damage in the brain: mechanisms and implications for neurodegenerative diseases. J Neurochem 68: 2227-40
110. Granger DN, Korthuis RJ (1995) Physiologic mechanisms of postischemic tissue injury. Ann Rev Physiol 57: 311-32
111. Sakuma S, Fujimoto Y, Sakamoto Y, Uchiyama T, Yoshioka K, Nishida H, Fujita T (1997) Peroxynitrite induces the conversion of xanthine dehydrogenase to oxidase in rabbit liver. Biochem Biophys Res Comm 230: 476-9

112. Szabó C, Szabó A, Cuzzocrea S, Zingarelli B, Salzman AL (1997) Evidence for the peroxynitrite - poly (ADP-ribose) synthetase pathway in the pathogenesis of mesenteric artery reperfusion injury. Jp J Pharmacol 75: (Suppl) 101P
113. Zingarelli B, Cuzzocrea S, Zsengeller, Z, Salzman AL, Szabó C (1997) Beneficial effect of inhibition of poly-ADP ribose synthetase activity in myocardial ischemia-reperfusion injury. Cardiovasc Res 36: 205-215
114. Wallis RA, Panizzon KL, Hanry D, Wasterlain CG (1993) Neuroprotection against nitric oxide injury with inhibitors of ADP-ribosylation. Neuroreport 5: 245-248
115. Zhang J, Pieper A, Snyder SH (1995) Poly(ADP-ribose) synthetase activation: an early indicator of neurotoxic DNA damage. J Neurochem 65: 1411-4
116. Cosi C, Suzuki H, Milani D, Facci L, Menegazzi M, Vantini G, Kanai Y, Skaper SD (1994) Poly(ADP-ribose) polymerase: early involvement in glutamate-induced neurotoxicity in cultured cerebellar granule cells. J Neurosci Res 39: 38-46
117. Endres M, Wang ZQ, Namura S, Waeber C, Moskowitz MA (1997) Ischemic brain injury is mediated by the activation of poly (ADP-ribose) synthetase. J Cerebral Blood Flow Metabol, in press
118. Kooy N, Royall J, Ischiropoulos H, Beckman J (1995) Peroxynitrite-mediated oxidation of dihydrorhodamine 123. Free Rad Biol Med 16:149-155
119. Eiserich JP, Cross CE, Jones AD, Halliwell B, Van der Vliet A (1996) Formation of nitrating and chlorinating species by reaction of nitrite with hypochlorous acid. A novel mechanism for nitric oxide-mediated protein modification. J Biol Chem 271: 19199-208
120. Van der Vliet A, Eiserich JP, Halliwell B, Cross CE (1997) Formation of reactive nitrogen species during peroxidase-catalyzed oxidation of nitrite. A potential additional mechanism of nitric oxide-dependent toxicity. J Biol Chem 272: 7617-25, 1997
121. Eiserich J P, Hristova M, Cross CE, Jones AD, Freeman BA, Halliwell B, Van der Vliet A (1997) Formation of nitric oxide dderivatives catalysed by myeloperoxidase in neutrophils. Nature, in press.
122. Szabó C, Salzman AL (1995) Endogenous peroxynitrite is involved in the inhibition of cellular respiration in immuno-stimulated J774.2 macrophages. Biochem Biophys Res Comm 209: 739-743
123. Hooper DC, Bagasra O, Marini JC, Zborek A, Ohnishi ST, Kean R, Champion JM, Sarker AB, Bobroski L, Farber JL, Akaike T, Maeda H (1997) Prevention of experimental allergic encephalomyelitis by targeting nitric oxide and peroxynitrite: implications for the treatment of multiple sclerosis. Proc Natl Acad Sci USA 94: 2528-33
124. Ames BN, Cathcart R, Schwiers E, Hochstein P (1981) Uric acid provides an antioxidant defense in humans against oxidant- and radical-caused aging and cancer: a hypothesis. Proc Natl Acad Sci USA 78: 6858-62
125. Wang ZQ, Auer B, Stingl L, Berghammer H, Haidacher D, Schweiger M, Wagner EF (1997) Mice lacking ADPRT and poly(ADP-ribosyl)ation develop normally but are susceptible to skin disease. Genes Develop 9: 509-20
126. DeMurcia JM, Niedergang C, Trucco C, Ricoul M, Dutrillaux B, Mark M, Oliver FJ, Masson M, Dietrch A, LeMeur M, Waltzinger C, Chambon P, DeMurcia G (1997) Requirement of poly (ADP-ribose) polymerase in recovery from DNA damage in mice and in cells. Proc Natl Acad Sci USA 94: 7303-7307

The Role of Nitric Oxide and Kinin on the Renal Water-Sodium Metabolism

Nobuyuki Ura, Yoshitoki Takagawa, Jun Agata, Kazuaki Shimamoto[1]

SUMARRY. Neutral endopeptidase 24.11 (NEP) catabolizes atrial natriuretic peptide (ANP) and kinin, and NEP inhibition results in diuresis and natriuresis. To further investigate the mechanisms of renal effects of NEP inhibitor (NEPI), we observed the role of nitric oxide (NO) together with kinin and ANP. NEPI, UK 73967 or thiorphan and kinin's receptor antagonist, Hoe 140 (Hoe) were employed, with or without a pretreatment of NO synthase inhibitor, N^{ω}-monomethyl-L-arginine (L-NMMA) in normal rats. Urinary kinin, NO^-_2+NO^-_3(NOx), cGMP, urine volume (UV) and urinary sodium excretion (UNaV) before and after NEPI, and plasma ANP level at the end of experiment, were evaluated. None of the variables changed with vehicle. There were significant increase in kinin, NOx, cGMP, UV and UNaV by NEPI. There were significant positive correlations between Δkinin and ΔUV or ΔUNaV, ΔNOx and ΔUV or UNaV, and ΔcGMP and ΔUNaV. However, there was no difference in plasma ANP between vehicle and NEPI groups. Hoe cancelled the increases of UV and UNaV caused by NEPI. With a pretreatment of L-NMMA, NEPI significantly increased kinin, while cGMP, UV and UNaV did not increase.

In conclusion, augmented renal kinin may play an important role in the renal water-sodium metabolism by NEPI, and renal NO may contribute to the kinin's action on this mechanism, while ANP may not contribute to it at least in normotensive rats. Moreover, changes in urinary cGMP does not reflect the changes in plasma ANP, but reflect the changes in renal NO under this condition.

KEY WORDS: neutral endopeptidase inhibitor, renal water-sodium metabolism, renal kinin, renal NO, urinary cGMP

---

[1] The Second Department of Internal Medicine, Sapporo Medical University School of Medicine, Sapporo 060-8543, Japan

Introduction

The role of the renal kallikrein-kinin system in the renal water-sodium metabolism [1-4] has been extensively reported as well as in the pathophysiology of various hypertensive diseases [5-8]. Kinin producing enzyme, kallikrein have mainly been examined to investigate the activity of this system, while there are some discrepancies between urinary kallikrein activity and the biologically active substances, kinin [9,10]. Renal kininases have received less attention, though they are one of the most important components in the regulation of intrarenal kinin. Neutral endopeptidase 24.11 (NEP) degrades kinin as well as kininase I and kininase II (angiotensin converting enzyme). Recently, we found the presence of NEP in the distal part of the nephron, the active site of the renal kallikrein-kinin system, in canine kidney [11], and others have reported the presence of NEP in cortical collecting duct and outer medullary collecting duct of rat kidney [12]. Additionally, urinary NEP contributes more than half of the renal kininases in the rat [13] and human [14]. Moreover, urinary NEP is significantly higher in patients with essential hypertension, primary aldosteronism or Cushing's syndrome than in normotensive subjects [15]. Accelerated renal NEP may play an important role in disorders of the renal water-sodium handling and in blood pressure elevation in these hypertensive diseases by disrupting the metabolism of intrarenal kinin.

Atrial natriuretic peptide [ANP] has been reported to regulate the water-sodium metabolism [16,17]. NEP also metabolizes and inactivates ANP [18] as well as ANP clearance receptors [19]. Many reports have indicated that the renal effects of NEP inhibitor depend on the inhibition of ANP metabolism [20-23], but some of them showed an increase in urinary cGMP, a second messenger of ANP [22,23], rather than the increase in plasma ANP.

It is well established that vascular kinin stimulate nitric oxide (NO) synthase activity, and part of the vasodilating action of kinin depends on NO [24,25]. However, it has not been clarified yet whether NO contributes to the effects of renal kinin. NO is also stimulated to biological activity by the induction of cGMP.

In this study, we attempted to evaluate the mechanisms of diuretic and natriuretic effects of NEP inhibitor by measuring the renal kinin, plasma ANP and NO. We also investigated the mechanism of increase in urinary cGMP without the increase in plasma ANP by NEP inhibitor, which was reported from other institute [22,23].

Materials and methods

Experiments were carried out on male Sprague-Dawley rats weighing 250 to 330 g. Food was withheld for 12 hours prior to study, but water was ingested ad libitum.

On the day of experiments, rats were anesthetized with intraperitoneal sodium pentobarbital (50 mg/kg) with supplemental doses (0.5 mg/kg/min) as necessary to maintain an even plane of anesthesia. A heating pad adjusted to 37°C was placed beneath the rats to maintain body temperature, and the trachea was cannulated to allow them to breathe spontaneously. A femoral artery was catheterized (PE-50, Clay-Adams, Parsippany, NJ) for monitoring the blood pressure and heart rate and for blood sampling. A femoral vein was catheterized (PE-50) for infusion of test solutions and for supplemental pentobarbital administration. After a midline lower abdominal incision, the bladder was isolated and catheterized with PE-90 tubing, taking special care to minimize trauma to the bladder. The presence of blood in urine was assessed by Hemacombistix (Ames, Miles-Sankyo, Tokyo); only rats showing no hematuria were included in these studies.

Soon after surgery, a saline (0.15 M NaCl) load equal to 5 % body weight was given intravenously at 20 ml/h, followed by a maintenance infusion (0.1 ml/min) of saline. After a 90 min equilibration interval, urine was collected over a 60-min control and then a 60-min experimental period for measurement of urine volume (UV), urinary sodium excretion (UNaV), kinin, NO^-_2+NO^-_3 (NOx) and cGMP. Urine samples were collected in preweighed polypropylene tubes at 0°C and 1 ml of each sample was transferred to another tube containing pepstatin (400 ng/10 µl; Peptide Institute Inc., Osaka) HCl for estimation of urinary kinin. Blood samples were taken at the end of the protocol and tested for plasma ANP.

Study 1): In group I (time control, n=8), saline was injected at a dose of 1 ml/kg intravenously and 0.5 ml/kg subcutaneously. In group II (n=9), specific NEP inhibitor, UK 73967 (UK; Pfizer Central Research, Sandwich, Kent) was injected in a dose of 10 mg/kg intravenously and 0.5 ml/kg of saline subcutaneously. In group III (n=10), UK was injected in the same amount as in group I and Hoe 140 (Hoe; Department of Pharma Synthesis, Hoechst AG, Frankfurt) was injected in a dose of 20 nmol/kg subcutaneously. All reagents and vehicle were injected at the beginning of the experimental period.

In group IV (n=8) and group V (n=8), the continuous infusion of N^{ω}-monomethyl-L-arginine (L-NMMA; Sigma Chemical Company, St. Louis, MO) in a dose of 200 µg/kg/min was started at 60 minutes prior to the control period, and the same procedures were performed as in groups I and II, respectively.

Study 2): In group I (time control, n=8), saline was injected at a dose of 1 ml/kg intravenously. In group II (n=10), specific NEP inhibitor, thiorphan (Shionogi Co., Ltd., Osaka) was injected in a dose of 30 mg/kg intravenously.

UV was determined gravimetrically, and UNaV, with ion electrode method (Ion Meter CIM-104A, Shimadzu Corp., Kyoto). Urinary kinin were measured by direct radioimmunoassay [26]. Urinary NOx was measured by an automated system

TCI-NOX 1000m (Tokyo Kasei Kogyo Co., Ltd., Tokyo) based on colormetric method [27]. Urinary cGMP was measured by direct radioimmunoassay (Cyclic GMP Assay Kit; Yamasa Shoyu, Chosi). Plasma ANP was extracted with a Sep-Pak C18 cartridge (Waters Chromatography Div., Millipore Corp., Milford, MA) and its concentration was measured by radioimmunoassay (Peninsula Laboratories. Belmont, CA).

All values were expressed as mean ± SEM. Statistical analysis was performed with Student's *t*-test and analysis of variance (ANOVA). A value of $p<0.05$ was considered significant.

Results

Study 1): Mean blood pressure, heart rate did not change in all groups. Urinary kinin (from 32 ± 8 to 36 ± 13 pg/kg/min), UV (from 300 ± 31 to 299 ± 32 μl/kg/min) and UNaV (from 43 ± 5 to 44 ± 4 μEq/kg/min) for the control group (group I) also showed no change during the experimental period, while cGMP (from 52 ± 3 to 39 ± 4 pmol/kg/min, $p<0.05$) decreased slightly but significantly. Following the injection of UK into group II, urinary kinin (from 44 ± 15 to 89 ± 25 pg/kg/min, $p<0.01$), cGMP (from 49 ± 3 to 71 ± 13 pmol/kg/min, $p<0.05$), UV (from 303 ± 37 to 439 ± 33 μl/kg/min, $p<0.01$) and UNaV (from 45 ± 4 to 63 ± 4 μEq/kg/min, $p<0.01$) increased significantly (Fig. 1). These variables were significantly higher in UK group than in the control group at the experimental period. There were significant positive correlations between Δkinin and ΔUV ($r=0.497$, $p<0.05$) or ΔUNaV ($r=0.576$, $p<0.05$), and between ΔcGMP and ΔUNaV ($r=0.501$, $p<0.05$). In group III, injected UK did not increase UV or UNaV with simultaneous administration of Hoe. Consequently, the changes of UV and UNaV were significantly higher in the UK group than those in the control group, while there was no significant difference in these changes between the control group and UK+Hoe group.

Mean blood pressure increased about 15 mmHg following L-NMMA treatment, but did not change during the control or experimental periods. In group IV (control group for group V), none of the variables changed significantly during the experimental period; in group V, injected UK significantly increased kinin (from 44 ± 7 to 73 ± 10 pg/kg/min, $p<0.05$) as seen in group II. However, urinary cGMP, UV and UNaV did not change by UK with a pretreatment of L-NMMA (Fig. 1).

There was not a significant difference in plasma ANP between the control group (61 ± 6 pg/ml) and UK group (77 ± 11 pg/ml), while urinary cGMP significantly increased in UK group and remained significantly higher than in the control group in the experimental period as mentioned above. Although groups IV (69 ± 7 pg/ml) and V (63 ± 7 pg/ml) also showed no difference in plasma ANP, urinary cGMP did not increase in group V, the UK group with a pretreatment of L-NMMA.

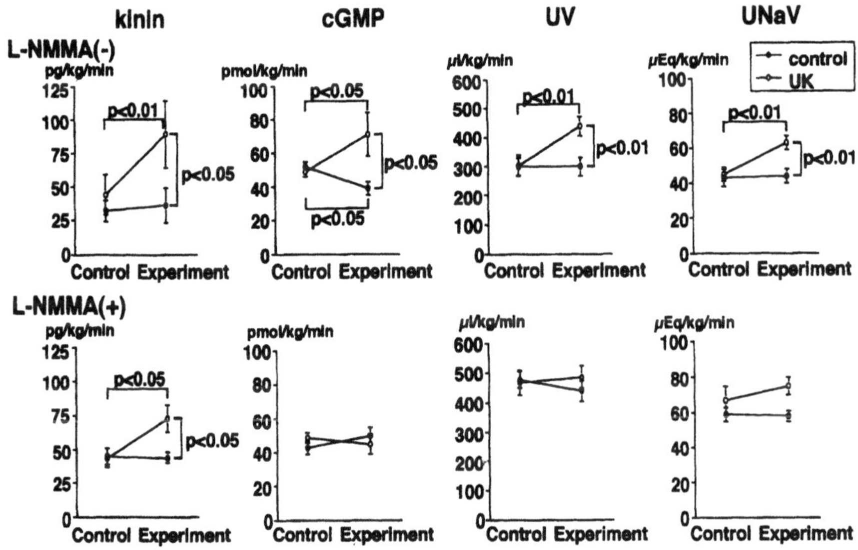

Fig. 1. Urinary excretion of kinin, cGMP, urine volume (UV) and urinary excretion of sodium (UNaV) before and after an injection of UK 73967 or vehicle with or without pretreatment of L-NMMA.

Study 2): Mean blood pressure, heart rate did not change in both groups. UV increased significantly at the experimental period in thiorphan group (group II; from 367 ± 29 to 432 ± 32 μl/kg/min, $p<0.01$), while it did not change significantly in the control group (group I; from 371 ± 49 to 326 ± 27 μl/kg/min). UV at the experimental period was significantly ($p<0.05$) higher in thiorphan group than in the control group. Although UNaV did not changed significantly at the experimental period in either the control group (from 43 ± 4 to 37 ± 3 μEq/kg/min) or thiorphan group (from 43 ± 4 to 50 ± 2 μEq/kg/min), it was significantly ($p<0.01$) higher in thiorphan group than in the control group at the experimental period.

There was a significant ($p<0.05$) increase in NOx at the experimental period by thiorphan injection (from 32 ± 2 to 41 ± 5 nmol/kg/min), while no change was found in the control group (from 30 ± 2 to 35 ± 3 nmol/kg/min).

Correlation between ΔNOx and ΔUV or ΔUNaV in both groups are given in Fig. 2. There were significant positive correlations between ΔNOx and ΔUV ($r=0.549$, $p<0.05$) or ΔUNaV ($r=0.488$, $p<0.05$).

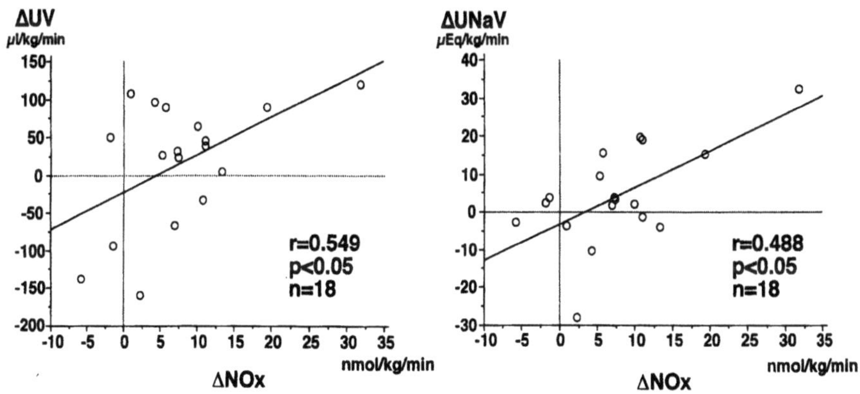

Fig. 2. Correlations between the changes in urinary excretion of NOx (ΔNOx) and those in urine volume (ΔUV) or urinary excretion of sodium (ΔUNaV) in thiorphan and vehicle treared groups.

Discussion

It is well established that NEP cleaves several kinds of peptides, kinin, ANP, enkephalins, endorphin, substance P and angiotensins in vitro. Kinin is inactivated by NEP mainly cracking at the Pro7-Phe8 bond, and ANP, at the Cys7-Phe8 bond. We have already showed that NEP is the main kininase in rat [13] and human urine [14]. The renal kininases are rich in the proximal part of the nephron, while a high amount of NEP also exists in the distal part of the nephron [11,12], which is known as the active site of the renal kallikrein-kinin system. Accordingly, renal NEP may play some role in regulation of intrarenal concentration of kinin. Recent publications [20-23] have indicated that the NEP inhibitor may work through inhibition of the ANP metabolism; on the other hand, some investigators have failed to show any increase in plasma ANP by infusion of NEP inhibitor [22,23]. It is still controversial whether acceleration of renal kinin is the main mechanism of renal effects of NEP inhibition or not.

Injected UK significantly increased excretion of urinary kinin, UV and UNaV without any changes in blood pressure and heart rate. Moreover, simultaneous administration of a specific kinin receptor antagonist Hoe canceled the renal effects of the NEP inhibitor. These observations strongly support the importance of renal kinin on the diuretic and natriuretic mechanisms of NEP inhibition. In other words, renal NEP may play an important role in sodium excretion by regulating the intrarenal concentration of kinin. However these data do not eliminate the possibility that mechanisms other than renal kinin may be involved, because of the incomplete suppression of natriuresis of NEP inhibitor by Hoe.

It is well known that kinin induce NO in vascular endothelial cells [28]. The vasodilating action of kinin is suspected by some to stem from an indirect mechanism, i.e., production of NO [24,25]; however, it has not been clarified yet whether NO contributes to the action of kinin in the kidney. Thereupon, the role of NO in the mechanism of renal action of kinin was investigated.

In this study, pretreatment of NO synthase inhibitor L-NMMA canceled the increases of UV and UNaV caused by NEP inhibitor, while NEP inhibitor still increased urinary kinin. We believe that this is the first demonstration that the renal action of kinin is mediated at least in part by NO in vivo. NEP activities both in plasma [29] and in vascular tissue [30] have been reported to be very low in normotensive rats. Moreover, NEP inhibitor does not change blood pressure, heart rate, renal blood flow or glomerular filtration rate in normotensive rats [13]. Taking these observations into consideration, NEP inhibitor may work through modulation of renal tubular reabsorption mechanism of luminal kinin and NO, rather than by changing renal hemodynamics. Stoos et al. [31] reported the renal action of luminal NO on electrolyte metabolism by using cultured cortical collecting tubular cell of mouse. Although the origin of luminal NO is still unclear, some reports [32,33] showed the existance of mRNA of the constitutive NO synthase in collecting tubules. It is not so unreasonable to suppose that renal kinin stimulate NO production in the collecting tubules, and that the two factors together regulate renal water-sodium handling. Recent studies [34,35] have shown the data concerning the mechanisms of the regulation of renal sodium metabolism of luminal NO.

In our experiment, plasma ANP did not change with NEP inhibitor. This study was carried out in the condition of moderate hypervolemia, and the production of plasma ANP increases by volume expantion. As UK inhibit the destruction of ANP, plasma levels of ANP may be overestimated, or at least not be underestimated by hypervolemia. Accordingly, hypervolemia may not be the reason for lack of an increase in plasma ANP. Prior to this study, we have performed some preliminary investigation to decide the protocol of this study. In preliminary study, urine samples were collected every 15 min for 60 min after UK administration, and UV, UNaV and urinary NEP activity were evaluated. UV and UNaV significantly increased and NEP decreased, and these were significantly differ from those in vehicle treated rats even at the final collection period. Although plasma ANP was not investigated in each period, NEP activity was suppressed for at least 60 min by bolus administration of UK. Bolus injection may not be the reason for lack of an increase in plasma ANP. The brush border of renal proximal tubules is rich in NEP. Although NEP inhibitor phosphoramidon inhibited 95 % of urinary NEP activity, infused synthetic bradykinin was not found in intact form in urine [13]. We have no data concerning urinary excretion of ANP, while NEP activity in the proximal tubules may still enough to inhibit the filtration of intact ANP to ANP receptor sites on the inner medullary collecting ducts by this amount of UK treatment. Moreover, UK did not increase urinary cGMP with a pretreatment of L-NMMA. If diuresis and natriuresis of UK is the result of focal increase of ANP in the distal part of the nephron, urinary cGMP should increase. ANP is metabolized by both NEP and ANP clearance receptors [18,19]. Because of the low levels of plasma [29] and vascular NEP [30] and the large contribution of clearance receptors to ANP metabolism in comparing NEP [36,37], plasma ANP might change little with NEP inhibition, at least in normotensive rats.

Although plasma ANP did not change with NEP inhibition, renal excretion of cGMP significantly increased. Moreover, increase in urinary cGMP was completely suppressed by the pretreatment of L-NMMA. Some recent publications [22,23] concluded that the renal action of NEP inhibitor depends on increased ANP, while they showed zero increase in endogeneous ANP in spite of a finite increase in urinary cGMP. From our results, it is very clear that the increase in urinary cGMP by NEP inhibition is mainly caused by NO, which is increased by augmented renal kinin and not by ANP under this condition.

In a human study, urinary NEP activity increased in patients with essential hypertension, primary aldosteronism and Cushing's syndrome [15]. Some reports have indicated that the administration of NEP inhibitor induced diuresis and natriuresis in normal subjects [20,23] and essential hypertensives[38,39]. Therefore, NEP inhibitor may have some potential for treating various hypertensive diseases.

In conclusion, augmented kinin may play an important role in the renal water-sodium metabolism by NEP inhibition, while ANP may not contribute to it at least in normotensive rats. Moreover, NO may participate in the renal effects of kinin induced by NEP inhibition. Changes in urinary cGMP follow changes in NO, rather than in plasma ANP under this condition.

References

1. Kauker ML (1980) Bradykinin action on the efflux of luminal 22Na in the rat nephron. J Pharmacol Exp Ther 214: 119-123
2. Tomita K, Pisano JJ, Knepper MA (1985) Control of sodium and potassium transport in the cortical collecting duct of the rat. -Effects of bradykinin, vasopressin, and deoxycorticosterone. J Clin Invest 76: 132-136

3. Tomiyama H, Scicli AG, Scicli GM, Carretero OA (1990) Renal effects of Fab fragments of kinin antibodies on deoxycorticosterone acetate-salt-treated rats. Hypertension 15: 413-414.
4. Shimamoto K, Ura N, Nakao T, Nishimiya T, Mita T, Kondo M, Ando T, Tanaka S, Iimura O (1983) Role of the renal kallikrein-kinin system in sodium metabolism in normotensives and essential hypertensives. New Zealand Med J 96: 905-907
5. Margolius HS, Geller R, Pisano JJ, Sjoerdsma A (1971) Altered urinary kallikrein excretion in human hypertension. Lancet ii: 1063-1065
6. Carretero OA, Scicli AG (1980) The renal kallikrein-kinin system. Am J Physiol 238: F247-F255
7. Ura N, Shimamoto K, Nakao T, Ogasawara A, Tanaka S, Mita T, Nishimiya T, Iimura O (1983) The excretion of human urinary kallikrein quantity and activity in normal and low renin subgroups of essential hypertension. Clin Exp Hypertens A5: 329-337
8. Iimura O, Shimamoto K, Ura N, Nakagawa M, Nishimiya T, Ando T, Yamaguchi Y, Masuda A, Ogata H, Saito S, Yamaji I, Fukuyama S (1987) The pathophysiological role of renal dopamine, kallikrein-kinin and prostaglandin systems in essential hypertension. Agents and actions 22: 247-256
9. Scicli AG, Rabito S, Carretero OA (1982) Blood and urinary kinins in human subjects during normal and low sodium intake. Adv Exp Med Biol 156A: 877-882
10. Ura N, Shimamoto K, Ogata H, Sakakibara T, Ando T, Fukuyama S, Nakagawa M, Saito S, Tanaka S, Iimura O (1989) The role of renal kininases in primary aldosteronism. Adv Exp Med Biol 247B: 145-150
11. Sakakibara T, Ura N, Shimamoto K, Ogata H, Ando T, Fukuyama S, Yamaguchi Y, Masuda A, Mori Y, Saito S, Ise T, Sasa Y, Yamauchi K, Iimura O (1989) Localization of neutral endopeptidase in the kidney determined by stop-flow method. Adv Exp Med Biol 247B: 349-353
12. Maeda Y, Tomita K, Ujiie K, Iino Y, Yoshiyama N, Shiigai T (1988) Renal function, hypertension and kallikrein-kinin system. Edited by Iimura O, Margolius HS. University of Tokyo Press, Tokyo, pp 99-102
13. Ura N, Carretero OA, Erdös EG (1987) Role of renal endopeptidase 24.11 in kinin metabolism in vitro and in vivo. Kidney Int 32: 507-513
14. Ogata H, Ura N, Shimamoto K, Sakakibara T, Ando T, Nishimiya T, Masuda A, Ise T, Shiiki M, Uno K, Iimura O (1989) A sensitive method for differential determination of kininase I, II and neutral endopeptidase (NEP) in human urine. Adv Exp Med Biol 247B: 343-348
15. Ura N, Shimamoto K, Satoh S, Kuroda S, Nomura N, Ohmoto Y, Masuda A, Iimura O (1993) Renal kininase I, kininase II and neutral endopeptidase 24.11 activities in patients with essential hypertension, primary aldosteronism and Cushing's syndrome. Hypertens Res 16: 253-258
16. Blaine EH, Seymour AA, Marsh EA, Napier MA (1986) Effects of atrial natriuretic factor on renal function and cyclic GMP production. Fed Proc 45: 2122-2127
17. Sonnenberg H, Honrath U, Chong CK, Wilson DR (1986) Atrial natriuretic factor inhibits sodium transport in medullary collecting duct. Am J Physiol 250: F963-F966
18. Stephenson SL, Kenny AJ (1987) The hydrolysis of alpha human atrial natriuretic peptide by pig kidney microvillar membranes is initiated by endopeptidase 24.11. Biochem J 243: 183-187

19. Maack T, Suzuki M, Almeida FA, Nussensweig D, Scarborough RM, McEnroe GA (1987) Physiological role of silent receptors of atrial natriurretic factor. Science 238: 675-678
20. Gros C, Souque A, Schwartz JC, Duchier J, Cournot A, Baumer P, Lecomte JM: Protection of atrial natriuretic factor against degradation (1989) Diuretic and natriuretic responses after in vivo inhibition of enkephalinase (EC 3.4.24.11) by acetorphan. Proc Natl Acad Sci 86: 7580-7584
21. Seymour AA, Fennell SA, Swerdel JN (1989) Potentiation of renal effects of atrial natriuretic factor-(99-126) by SQ 29072. Hypertension 14: 87-97
22. Sybertz EJ, Chiu PJS, Vemulapalli S, Pitts B, Foster CJ, Watkins RW, Barnett A (1989) SCH 39370, a neutral metalloendopeptidase inhibitor, potentiates biological responses to atrial natriuretic factor and lowers blood pressure in desoxycorticosterone acetate-sodium hypertensive rats. J Pharmacol Exp Ther 250: 624-631
23. Richards AM, Wittert G, Espiner EA, Yandel TG, Frampton C, Ikram H: Prolonged inhibition of endopeptidase 24.11 in normal man (1991) renal, endocrine and haemodynamic effects. J Hypertens 9: 955-962
24. Boulanger C, Schini VB, Moncada S, Vanhoutte PM (1990) Stimulation of cyclic GMP production in cultured porcine endothelial cells by bradykinin, adenosine diphosphate, calcium ionophore A23187 and nitric oxide. Br J Pharmacol 101: 152-160
25. Lahera V, Salom MG, Fiksen-Olsen MJ, Romero JC (1991) Mediatory role of endothelium-derived nitric oxide in renal vasodilatory and excretory effects of bradykinin. Am J Hypertens 4: 260-262
26. Shimamoto K, Ando T, Nakao T, Sakuma M, Miyahara M (1978) A sensitive radioimmunoassay method for urinary kinins in man. J Lab Clin Med 91: 721-728
27. Green LC, Wagner DA, Glogoeski J, Skipper PL, Wishnok IS, Tannenbaum SR (1982) Analysis of nitrate, nitrite, and [15N]niyrate in biological fluids. Anal Biochem 126: 131-138.
28. Palmer RMJ, Ferrige AG, Moncada S (1987) Nitric oxide release accounts for the biological activity of endothelium-derived relaxing factor. Nature 327: 524-526
29. Ishida H, Scicli AG, Carretero OA (1989) Role of angiotensin converting enzyme and other peptidases in in vivo metabolism of kinins. Hypertension 14: 322-327
30. Tamburini PP, Koehn JA, Gilligan JP, Charles D, Palmesino RA, SharifR, McMartin C, Erion MD, Miller MJS (1989) Rat vascular tissue contains a neutral endopeptidase capable of degrading atrial natriuretic peptide. J Pharmacol Exp Ther 251: 956-961
31. Stoos BA, Carretero OA, Farhy RD, Scicli G, Garvin JL (1992) Endotherium-derived relaxing factor inhibits transport and increases cGMP content in cultured mouse cortical collecting duct cells. J Clin Invest 89: 761-765
32. Terada Y, Tomita K, Nonoguchi H, Marumo F (1992) Polymerase chain reaction localization of constitutive nitric oxide synthase and soluble guanylate cyclase messenger RNAs in microdissected rat nephron segments. J Clin Invest 90: 659-665
33. Ujiie k, Yuen J, Hogarth L, Danziger R, Star RA (1994) Localization and regulation if endothelial NO synthase mRNA expression in rat kidney. Am J Physiol 269: F296-302
34. McKee M, Scavone C, Nathanson JA (1994) Nitric oxide, cGMP, and hormone regulaion of active sodium transport. Proc Natl Acad Sci 91: 12056-12060
35. Stoos BA, Garcia NH, Garvin JL (1995) Nitric oxide inhibits sodium reabsorption in the isolated perfused cortical collection duct. J Am Soc Nephrol 6: 89-94

36. Chiu PJS, Tetzloff G, Romano MT, Foster CJ, Sybertz EJ (1991) Influence of C-ANF receptor and neutral endopeptidase on pharmacokinetics of ANF in rats. Am J Physiol 260: R208-216
37. Chevalier RL, Garmey M, Scarborough RM, Linden J, Gomez RA, Peach MJ, Carey RM (1991) Inhibition of ANP clearance receptors and endopeptidase 24.11 in maturing rats. Am J Physiol 260: R1218-1228
38. Singer DRJ, Markandu ND, Buckley MG, Miller MA, Sagnella GA, MacGregor GA (1991) Dietary sodium and inhibition of neutral endopeptidase 24.11 in essential hypertension. Hypertension 18: 798-804
39. Richards AM, Crozier IG, Kosoglou T, Rallings M, Espiner EA, Nicholls MG, Yandel TG, Ikram H, Frampton C (1993) Endopeptidase 24.11 inhibition by SCH 42495 in essential hypertension. Hypertension 22: 119-126

Vascular Activities of Hemoglobin-Based Oxygen Carriers: Relationship Between Vasoconstrictive Activity and Endothelial Permeability

Kunihiko Nakai [1], *Ichiro Sakuma* [2], *Hiroshi Satoh* [1], *Akira Kitabatake* [2]

SUMMARY.Acellular hemoglobin (Hb)-based oxygen carriers have been studied for the substitution of red blood cell transfusion. However, they have been shown to induce several undesired reactions such as hypertension, gastrointestinal symptoms and platelet activation. Hb-induced scavenging of nitric oxides might be responsible for the backgrounds of these reactions. Based on several in vitro experiments, we propose a hypothesis that acellular Hb derivatives having smaller molecular masses have increased endothelial permeabilities and therefore induce more potent vasoconstriction as a result of abluminal nitric oxide scavenging.

KEY WORDS: Blood substitutes, Endothelial permeability, Hemoglobin, Nitric oxide, Vasoconstriction

Introduction

The major concerns in the development of Hb-based oxygen carriers (HBOCs) have been the modulation of increased oxygen affinity and the short circulatory retention of native Hb. These problems have been resolved successfully by strategic chemical

---

1 Environmental Health Sciences, Tohoku University Graduate School of Medicine, Aoba- ku, Sendai, 980-0865 Japan
2 Department of Cardiovascular Medicine, Hokkaido University School of Medicine, Kita-ku, Sapporo, 060-8638 Japan

and genetic modifications of the protein, and clinical trials using some acellular Hb derivatives have already started (reviewed in Tsuchida [1]). Such modification methods include intramolecular and intermolecular crosslinking, conjugation to an inert material such as polyethylene glycol (PEG), and polymerization. However, it has been recently shown that the infusion of acellular Hb derivatives causes hypertension [2][3], gastrointestinal symptoms [4], and platelet activation [5]. In this report, we focus on the vascular effects of HBOCs.

One of the early reports that noted Hb-related hypertension was by Bayliss [6] in 1920, who observed a slight increase in blood pressure after the infusion of filtered hemolysate into shocked animals. From these early experiments, the basic question arose as to whether Hb itself is toxic, since hemolysate may contains unknown vasoconstrictive contaminants. Recently, Vogel et al. [7] provided confirmative data indicating the vasoconstrictive activity of Hb in isolated rabbit hearts. They also reported that polymerized pyridoxalated Hb still had a reduced vasoconstrictive activity. Indeed, it is not surprising that acellular Hb derivatives have vasoconstrictive activity. Endothelium-derived relaxing factor (EDRF) has recently been identified as nitric oxide (NO) or an NO-containing moiety [8], and Hb binds NO in two ways; the heme iron of Hb irreversibly binds NO with an extremely great affinity. The sulfhydryl groups of β93 cysteine of the protein also form a stable complex with NO [9]. This suggests that Hb itself is the major factor responsible for the observed vasoconstriction.

However, some puzzles still remain. The reason why a very small amount of acellular Hb induces significant vasoconstriction is unclear because peripheral blood already contains a large amount of cellular Hb in red cells. At least, it seems unlikely that acellular Hb-induced vasoconstriction is produced solely in the same way that red cells inhibit EDRF. The molecular mass of acellular Hb seems to have a significant impact on its vasoconstrictive activity because PEG-modification significantly reduced the vasoconstrictive property of diaspirin cross-linked Hb (DCLHb) [10]. The encapsulation of Hb into liposomes also reduced vasoconstriction in a coronary perfusion experiment [11] and in vivo [12]. It is clear that a more detailed understanding of the underlying mechanism(s) by which Hb induces vasoconstriction is necessary to ensure the safety of HBOCs. We have performed several in vitro experiments in an attempt to achieve this goal.

Organ bath experiments

The preparation of human and bovine stroma-free Hbs and several acellular Hb derivatives were described previously [13]. PEG-modified pyridoxalated human Hb (PEG-Hb) was a generous gift from Dr. Y. Iwashita (Ajinomoto Co. Ltd., Kawasaki, Japan). Intramolecularly crosslinked Hb (XL-Hb) was prepared from human SFH with bis(sulfosuccinimidyl)suberate [14]. Two liposome-Hb preparations, a liposome stabilized by membrane phospholipid polymerization

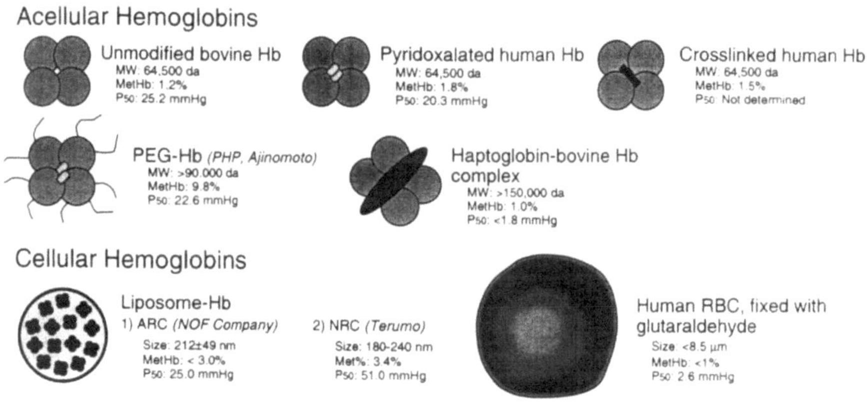

Fig. 1. Physicochemical characteristics of Hb preparations used in this report.

(ARC)[15] and a PEG-modified liposome (NRC)[16], were provided by NOF Company (Tsukuba, Japan) and Terumo (Kanagawa, Japan), respectively. Their physicochemical characteristics are illustrated in Fig. 1.

We have characterized Hb-induced vasoconstriction in an organ bath system using rabbit aortic strips [13]. In this in vitro experiment, all Hb preparations reversed acetylcholine (ACh)-induced relaxation in a dose-dependent manner. These observations were already reported by others [17]. On the other hand, we also noticed an interesting phenomenon in that the repeated exposure of tissues to unmodified Hb gradually reduced the responsiveness to ACh. We therefore performed an experiment: the tissues were pretreated with several Hb preparations for 30 min and then ACh-induced relaxation was recorded in the absence of Hb. Exposure to bovine SFH induced a significant reduction of the following ACh-induced relaxation, while the relaxation was either marginally reduced or not affected by haptoglobin-hemoglobin complex (Hp-Hb), PEG-Hb, liposome-Hb (ARC) and glutaraldehyde-fixed RBCs (Figs. 2 and 3). This persistent inhibitory effect by unmodified Hb disappeared when the tissue was further incubated for 2 - 3 hr with changed medium, indicating the reversibility of this inhibitory action and the functional integrity of the endothelium. Although vasoconstrictive factor(s) other than Hb might have contributed somewhat, this would not be expected to cause a significant effect, since Hp-Hb also contained the same components as unmodified Hb.

We performed similar experiments in the presence of 2% BSA or 2% dextran (mean MW 60,000 - 90,000 da). As shown in Fig. 3, both macromolecules partially prevented Hb-induced vasoconstriction.

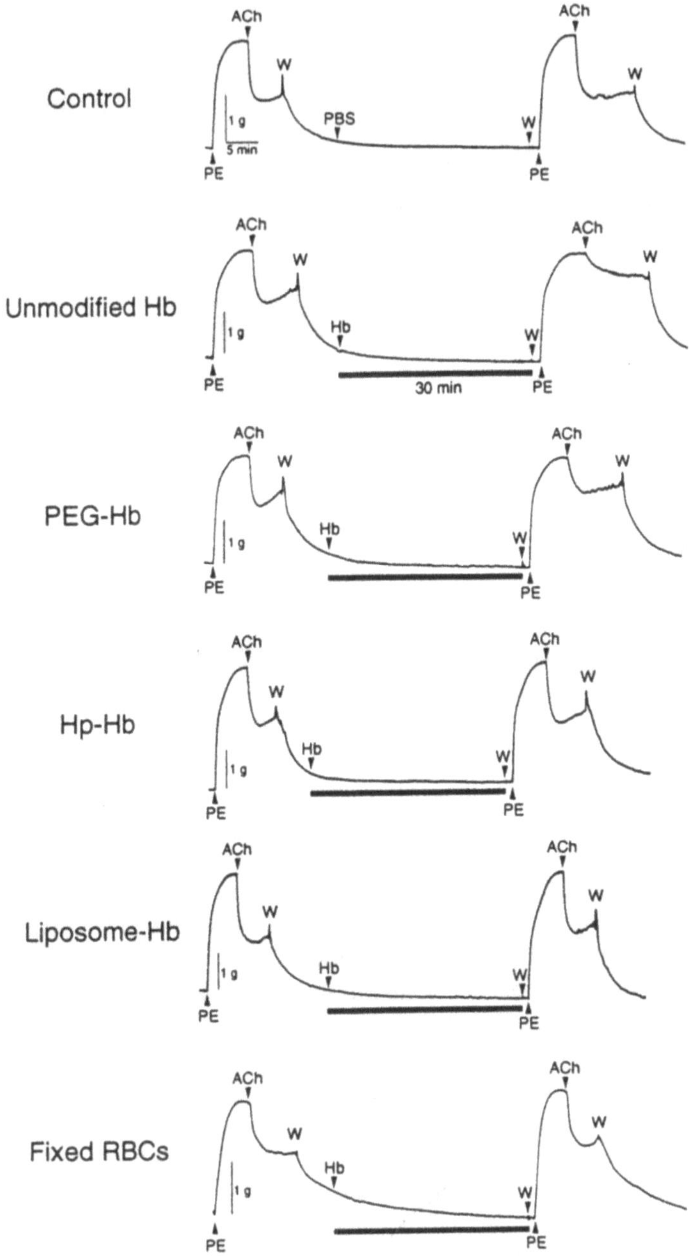

Fig. 2. Typical tracings showing inhibition of ACh-induced relaxation by several Hb preparations (Hb at 0.01%) in the organ bath assays using rabbit aortic strips. PE, phenylephrine 1 μM; ACh, acetylcholine 1 μM; W, wash-out.

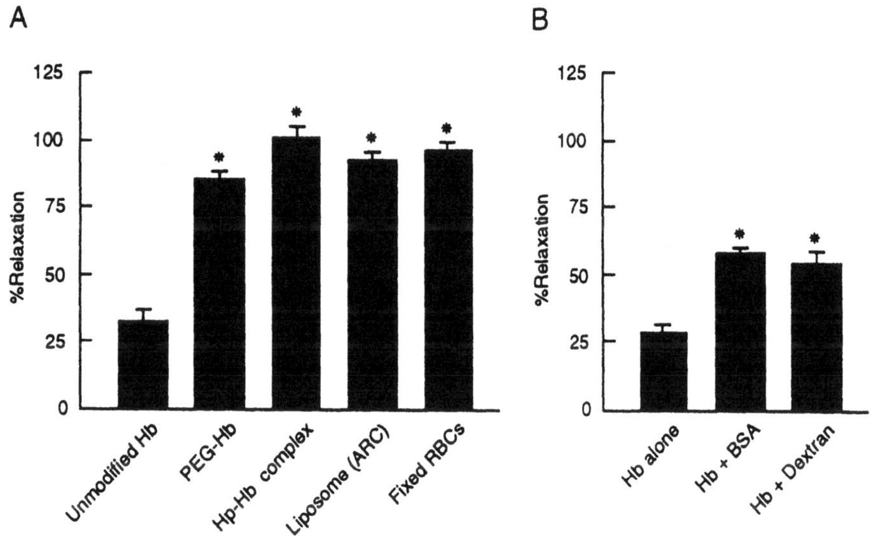

Fig. 3. Inhibition of ACh-induced relaxation by Hbs. A, Inhibitory effect of several Hb preparations (Hb at 0.01%). B, Unmodified Hb-induced inhibition in the presence of 2% BSA or 2% dextran. Means±SE of 5-12 experiments. *$p<0.01$ as compared with unmodified Hb.

These observations led us to propose the hypothesis that Hb-induced vasoconstriction needs infiltration of Hb into the tissues, probably through the intercellular cleft of the endothelium. It is likely that macromolecules such as albumin and dextran compete with Hb for access to the tissue.

Heart perfusion experiments

We have also examined the vascular activity of Hb preparations in a Langendorff perfusion model of rat hearts [11]. Changes in perfusion pressure and endothelium-dependent vasorelaxation, which were induced by a bolus injection of bradykinin (BK), were monitored.

Perfusion of rat hearts with bovine SFH and PEG-Hb increased perfusion pressure in a dose-dependent manner, and typical tracings are shown in Fig. 4. Unmodified bovine Hb increased coronary perfusion pressure and decreased the duration of BK-induced relaxation (Fig. 5). PEG-Hb also presented a similar vasoconstrictive profile with a reduced extent. In contrast, liposome-Hb (NRC) did not cause any vasoconstrictive response. We also observed that another liposome-Hb, ARC, had no vasoconstrictive activity [11].

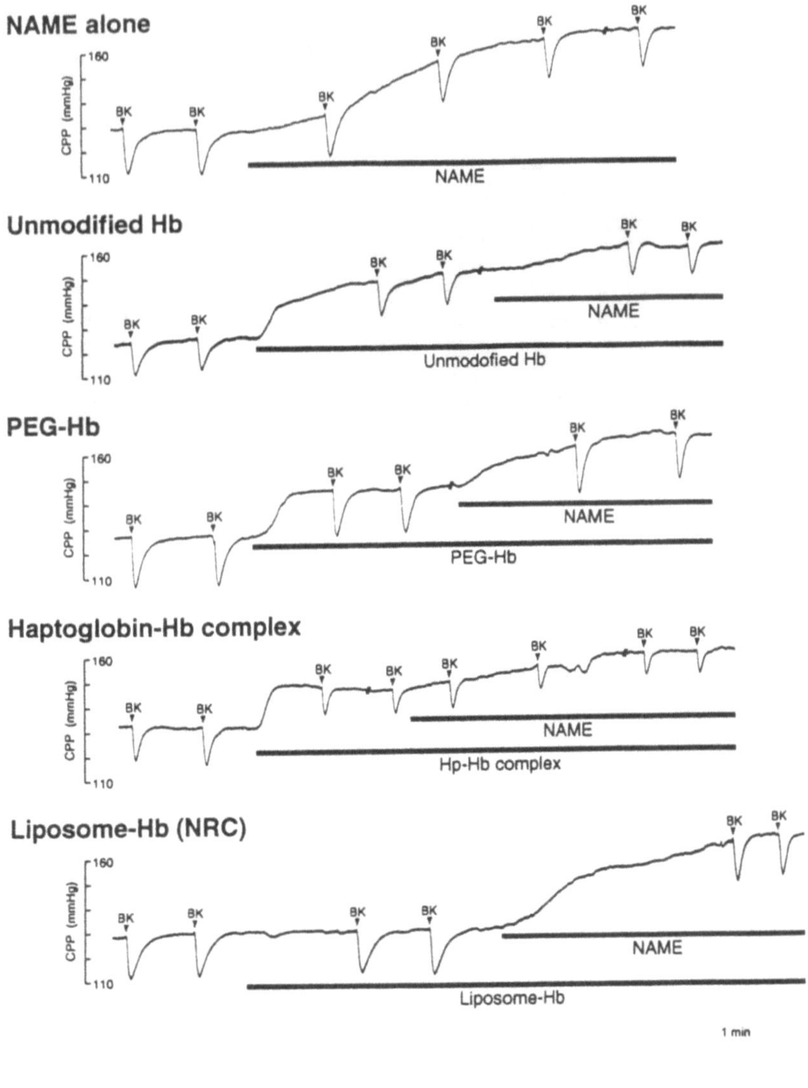

Fig. 4. Typical tracings showing Hb (0.1%)-induced vasoconstriction in rat perfused hearts. Hearts prepared from male Wister rats were perfused with Krebs-Henseleit solution containing 10 nM U-46619 and 0.2% BSA. Perfusion flow rate was adjusted as to maintain the coronary perfusion pressure (CPP) at 120 - 140 mmHg. BK, bradykinin 10 pmol in 4 μL; NAME, N^{ω}-nitro-L-arginine methyester 100 μM.

Both SFH and PEG-Hb affected only the duration of BK-induced relaxation, while neither none of them affected the amplitude of the relaxation. These vasoconstrictive profiles of acellular Hbs were in good agreement with the report of

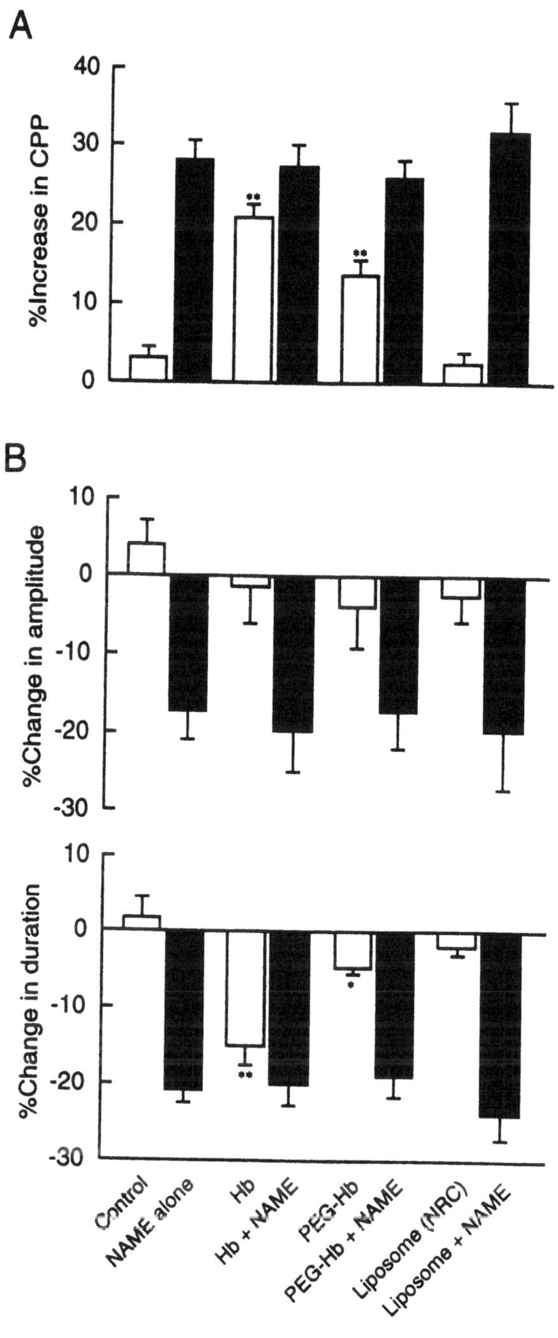

Fig. 5. Effects of Hb preparations on CPP and BK-induced relaxation. A, Changes in CPP B, Changes in amplitude and duration of BK-induced relaxation. $^*p<0.05$, $^{**}p<0.01$ as compared with control. Mean±SE of 8-10 experiments.

Sakuma et al. [18] who showed Hb-induced inhibitory effects on endothelin-induced vasodilation. BK-induced relaxation is caused by a combination of at least two factors, EDRF and endothelium-derived hyperpolarizing factor (EDHF): EDRF may relate mainly to the duration of BK-induced relaxation, while EDHF may mainly relate to the amplitude of the relaxation. Therefore, the profiles of BK-induced relaxation in the presence of Hb suggested that Hb disturbed only the action of EDRF. This point was also confirmed by the addition of a specific inhibitor for NO synthase, L-N^G-nitroarginine methylester (NAME), to the perfusate. NAME induced further vasoconstrictive effects in the Hb-induced response, but the extent of changes was identical for all conditions.

We also performed this perfusion experiment in the presence of BSA (0 - 2%) in the perfusate. BSA significantly delayed the acellular Hb-induced vasoconstrictive response in a dose-dependent fashion, while the extent of increase in perfusion pressure attained was not affected. Macromolecules such as BSA may inhibit Hb extravasation competitively.

Endothelium permeability experiments

Endothelial permeability characteristics of Hb preparations should be closely related to their vasoconstrictive activities if Hb induces vasoconstriction as a result of abluminal EDRF scavenging. Thus, we measured the transendothelial fluxes of several Hb preparations using a bovine thoracic aortic endothelial cell monolayer cultured on a microporous membrane as shown in Fig. 6 [19]. With medium placed in both the inner and outer wells, two chambers that were separated by the endothelial monolayer grown on the collagen filter were created. The data obtained were expressed in terms of the permeability coefficient (expressed in cm/s).

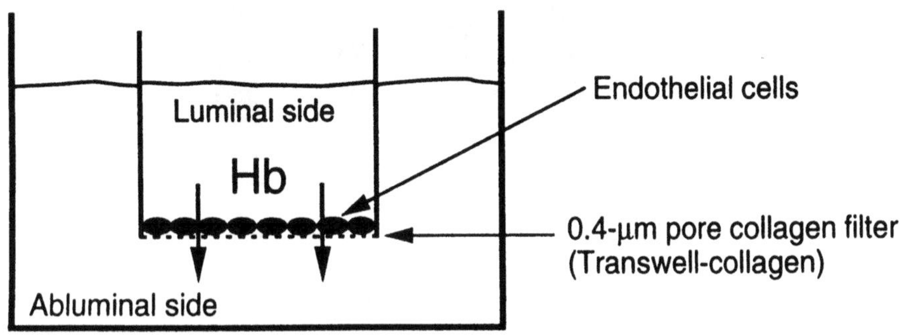

Fig. 6. Permeability measurement using cultured bovine endothelial cell monolayer. Bovine thoracic aortic endothelial cells were grown to confluence on culture plate inserts with a porous collagen filter.

Fig. 7 shows the permeability characteristics of Hb derivatives. With the untreated monolayer, the permeability coefficient of unmodified Hb was almost twice that of BSA. Intramolecular crosslinking slightly but significantly decreased the permeability of Hb. Other Hb modifications such as PEG modification and Hp binding profoundly reduced the transendothelial flux of Hb. Liposome-Hb (ARC) showed an even lower permeability coefficient. These results demonstrated for the first time that Hb derivatives having smaller molecular masses had increased permeabilities. This supports the idea that molecular mass is an essential factor contributing to the permeability characteristics, although we must consider other factors, including molecular shape and net charge, to obtain a precise understanding of decreased permeabilities of modified Hbs.

Endothelial barrier functions deteriorate in some pathological situations. In particular, hemorrhagic shock occurring after trauma, which is the most likely situation for the application of HBOCs, is accompanied by decreased endothelial permeability, in which IL-6 is a major factor regulating the permeability. Oxygen carriers would also likely be infused into trauma patients with concomitant endotoxemia, in which LPS disrupts the endothelial barrier function. LPS stimulates monocytes and macrophages to produce several pro-inflammatory cytokines, including IL-6, and these cytokines increase vascular endothelial permeability. Therefore, we examined the transendothelial Hb fluxes with endothelial monolayers pretreated with either IL-6 (100 ng/mL for 21 hr) or LPS (1 μg/ml for 10 hr). The results clearly indicated that acellular Hb derivatives moved

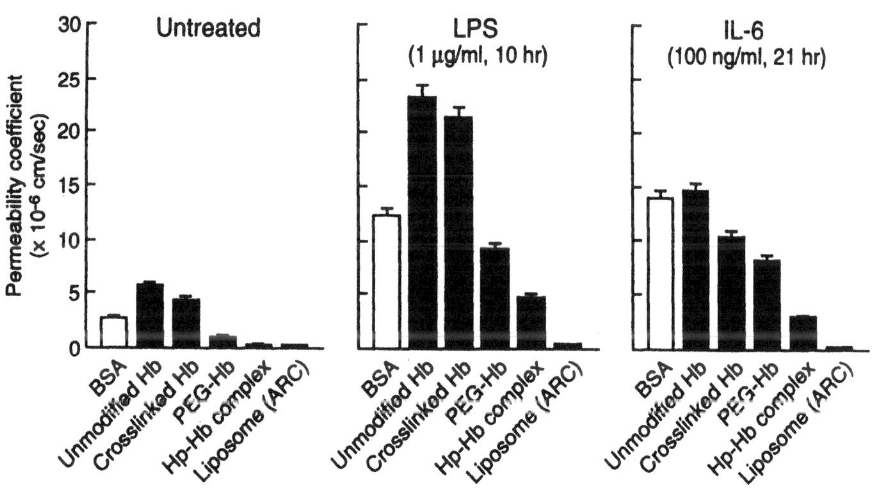

Fig. 7. Permeability characteristics of several Hb preparations in untreated, LPS (1 μg/mL, 10 hr)-treated or IL-6 (100 ng/mL, 21 hr)-treated monolayers. Means±SE of 5 to 10 experiments.

more rapidly in settings for the above pathophysiological situations in comparison with untreated monolayers. Interestingly, liposome-Hb showed little endothelial flux even after the pretreatment with IL-6 or LPS. Liposomes having a mean diameter of 200 nm appear to be large enough to prevent extravasation even under pathophysiological conditions.

Discussion

We here provide the first information demonstrating that the extravasation of Hb molecules into the vascular bed is a prerequisite for Hb-induced vasoconstriction, as summarized in Fig. 8. Extravasation could be inhibited by increasing the molecular mass of Hb derivatives, and, therefore, such larger products should have smaller vasoconstrictive activity. Although the exact pathway for extravasation is still unknown, we hypothesize that the endothelial intercellular cleft is a probable route.

Hypertensive reaction due to the vasoconstriction caused by acellular Hb derivatives is one of the unsolved problems in HBOCs. This effect has been shown to be long-lasting, but the extent of hypertension seems to be self-limiting. Therefore, this is unlikely to be a serious limitation. Far from that, hypertension itself may improve the hypotensive reaction during hemorrhagic shock [20]. DCLHb [21] and PEG-Hb [22] will be very effective to eliminate the excessive NO in sepsis. The blood flow in the brain and heart conversely increases during Hb-induced hypertension, while it decreases in other organs [23], presumably because

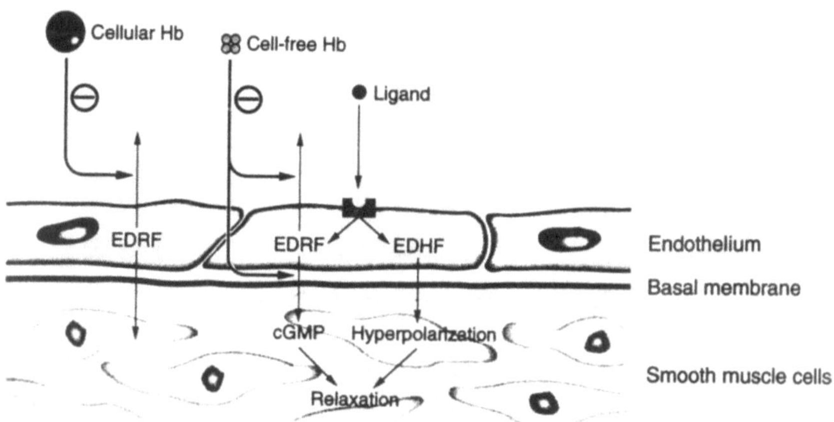

Fig. 8. Possible mechanism of acellular Hb-induced vasoconstriction as a result of abluminal EDRF scavenging.

EDHF is the predominant vasodilating factor in the brain and heart (Sakuma I. et al. unpublished observation). The presence of the blood-brain barrier also contributes to prevention of Hb extravasation in the cerebral circulation. Therefore, this differential vasoconstrictive activity can be utilized for the treatment of cerebral ischemic injury. On the other hand, hypertension due to the vasoconstriction of arterioles themselves may impair the oxygen supply to the tissues by reducing the blood flow downstream. It should be again pointed out that the extravasation of Hb might be a prerequisite for the vasoconstrictive reaction. Hb extravasated in turn may induce endothelial damage due to the heme-related oxidative-stress response [24]. The endothelial barrier function should decrease under some patho-physiological conditions; post-ischemic edema in cerebral circulation is closely related to the changes in permeability of the blood-brain barrier. Winslow [25] commented that whether vasoactivity will limit application of acellular HBOCs to clinical use is still open. Thus, the lack of full understanding regarding the consequences of Hb extravasation might be of importance.

Vasoactivity caused by HBOCs does not have a single etiology. Hb may potentiate an adrenergic mechanism [26]. There is some evidence of Hb involvement in the synthesis of endothelin-1 [27], although this topic is controversial [28]. Hb solution was found to initiate the synthesis of prostaglandins [29]. It has been emphasized that the extravasation of Hb is likely essential even in these hypothetical mechanisms.

Finally, other Hb-related side effects such as gastrointestinal symptoms and platelet activation may also be related to the removal of NO by Hb. NO plays a significant role in the nonadrenergic noncholinergic nerve-mediated relaxation of gastrointestinal smooth muscle [4]. Endothelium-released NO inhibits platelet function through increasing levels of platelet cGMP. These toxic effects are likely related to the EDRF scavenging by extravasated Hb.

Conclusion

Endothelial permeability of Hb derivatives is a very important factor relating to the difficulties observed with acellular Hb derivatives, especially to Hb-induced vasoconstriction. Endothelial permeability also has a great impact on its circulatory plasma retention. Thus, safety and efficacy issues are very closely related. In this context, it must be emphasized that the target of acellular Hb derivatives is the vascular endothelium. Successful clinical utilization of HBOCs will require a greater understanding of the interaction between Hbs and endothelium and the consequences of extravasation of Hb in both physiological and pathological situations. We believe that it is possible to engineer acellular HBOCs in order to improve biocompatibility. Our observations also highlight the significance of the cellularity of HBOCs. Great hopes are now held for the development of cellular HBOCs.

ACKNOWLEDGMENTS: We thank Drs. T. Ohta and Y. Nakazato for their helpful advice. Part of this study was supported by research grants from the Ministry of Education, Science and Culture (No. 09557125) and the Ministry of Health and Welfare of Japan.

References

1. Tsuchida E (1995) Introduction: Overview and Prospectives. Introduction: Overview and Prospectives. In: Tsuchida E (ed) Artificial Red Cells. Materials, Performances and Clinical Study as Blood Substitutes. John Wiley & Sons Ltd, Chichester, pp 1-20
2. Keipert PE, Gonzales A, Gomez CL, MacDonald VW, Hess JR, Winslow RM (1993) Acute changes in systemic blood pressure and urine output of conscious rats following exchange transfusion with diaspirin-crosslinked hemoglobin solution. Transfusion 33:701-708
3. Gulati A, Sharma AC, Burhop KE (1994) Effect of stroma-free hemoglobin and diaspirin cross-linked hemoglobin on the regional circulation and systemic hemodynamics. Life Sci 55:827-837
4. Rattan S, Rosenthal GJ, Chakder S (1995) Human recombinant hemoglobin (rHb1.1) inhibits nonadrenergic noncholinergic (NANC) nerve-mediated relaxation of internal anal sphincter. J Pharmacol Exp Ther 272:1211-1216
5. Olsen SB, Tang DB, Jackson MR, Gomez ER, Ayala B, Alving BM (1996) Enhancement of platelet deposition by cross-linked hemoglobin in a rat carotid endarterectomy model. Circulation 93:327-332
6. Bayliss WM (1920) Is haemolysed blood toxic? Br J Exp Pathol 1:1-9
7. Vogel WM, Dennis RC, Cassidy G, Apstein CS, Valeri CR (1986) Coronary constrictor effect of stroma-free hemoglobin solutions. Am J Physiol 251:H413-H420
8. Palmer RMJ, Ferrige AG, Moncada S (1987) Nitric oxide release accounts for the biological activity of endothelium-derived relaxing factor. Nature 327:524-526
9. Jia L, Bonaventura C, Bonaventura J, Stamler JS (1996) S-nitrosohaemoglobin: a dynamic activity of blood involved in vascular control. Nature 380:221-226
10. Nolte D, Botzlar A, Pickelmann S, Bouskela E, Messmer K (1997) Effects of diaspirin-cross-linked hemoglobin (DCLHbTM) on the microcirculation of striated skin muscle in the hamster: A study on safety and toxicity. J Lab Clin Med 30:314-327
11. Nakai K, Usuba A, Ohta T, Kuwabara M, Nakazato Y, Motoki R, Takahashi TA (1998) Coronary vascular bed perfusion with a polyethylene glycol-modified hemoglobin-encapsulated liposome, Neo Red Cell, in rats. Artif Organs (in press)
12. Rabinovici R, Rudolph AS, Vernick J, Feuerstein G (1994) Lyophilized liposome encapsulated hemoglobin: evaluation of hemodynamic, biochemical, and hematologic responses. Crit Care Med 22:480-485
13. Nakai K, Ohta T, Sakuma I, Akama K, Kobayashi Y, Tokuyama S, Kitabatake A, Nakazato Y, Takahashi TA, Sekiguchi S (1996) Inhibition of endothelium-dependent

relaxation by hemoglobin in rabbit aortic strips: Comparison between acellular hemoglobin derivatives and cellular hemoglobins. J Cardiovasc Pharmacol 28:115-123
14. Manjula BN, Roy RP, Smith PK, Acharya AS (1994) Bissulfosuccinimidyl esters of aliphatic dicarboxylic acids: a new class of 'affinity directed' beta beta crosslinkers of HbA. Artif Cells Blood Substit Immobil Biotechnol 22:747-752
15. Satoh T, Kobayashi K, Sekiguchi S, Tsuchida E (1992) Characteristics of artificial red cells hemoglobin encapsulated in poly-lipid vesicles. ASAIO J 38:M580-584
16. Takahashi A (1995) Characterization of Neo Red Cells (NRCs), their function and safety in vivo tests. Artif Cells Blood Substit Immobil Biotechnol 23:247-254
17. Martin W, Villani GM, Jothianandan D, Furchgott RF (1985) Selective blockade of endothelium-dependent and glyceryl trinitrate-induced relaxation by hemoglobin and by methylene blue in the rabbit aorta. J Pharmacol Exp Ther 232:708-716
18. Sakuma I, Asajima H, Fukao M, Tohse N, Tamura M, Kitabatake A (1993) Possible contribution of potassium channels to the endothelin-induced dilation of rat coronary vascular beds. J Cardiovasc Pharmacol 22(Suppl. 8):S232-234
19. Nakai K, Sakuma I, Ohta T, Ando J, Kitabatake A, Nakazato Y, Takahashi T (1998) Permeability characteristics of hemoglobin derivatives across cultured endothelial cell monolayers. J Lab Clin Med (in press)
20. DeAngeles DA, Scott AM, McGrath AM, Korent VA, Rodenkirch LA, Conhaim RL, Harms BA (1997) Resuscitation from hemorrhagic shock with diaspirin crosslinked hemoglobin, blood, or hetastarch. J Trauma 42:406-414
21. Burhop K, Ince C, Nolte D, Gulati A, Sibbald W, Malcolm D (1997) Overview of the effects of diaspirin crosslinked hemoglobin (DCLHb) on oxygenation and perfusion of the microcirculation. Abstract, VII International Symposium on Blood Substitutes (7-ISBS) 43
22. Bone HG, Nishida K, Booke M, McGuire R, Trader L, Trader D (1995) Effects of pyridoxalated hemoglobin-polyethylene conjugate (PHP) as an oxygen carrier. Shock (Supple)4:35
23. Waschke K, Schlock H, Albert DM, Van-Ackern K, Kuschinsky W (1993) Local cerebral blood flow and glucose utilization after blood exchange with a hemoglobin-based O2 carrier in conscious rats. Am J Physiol 265:H1243-1248
24. Motterlini R, Foresti R, Vandegriff K, Intaglietta M, Winslow R (1995) Oxidative-stress response in vascular endothelial cells exposed to acellular hemoglobin solutions. Am J Physiol 269:H648-H655
25. Winslow RM (1995) Hemoglobin-based red cell substitutes: unsolved issues and future directions. Hemoglobin-based red cell substitutes: unsolved issues and future directions. In: Tsuchida E (ed) Artificial Red Cells. John Wiley & Sons Ltd, New York, pp 117-130
26. Gulati A, Rebello S (1994) Role of adrenergic mechanisms in the pressor effect of diaspirin cross-linked hemoglobin. J Lab Clin Med 124:125-133
27. Gulati A, Sen AP, Sharma AC, Singh G (1997) Role of ET and NO in resuscitative effect of diaspirin cross-linked hemoglobin after hemorrhage in rat. Am J Physiol 273:H827-H836
28. Tai J, Kim HW, Greenburg AG (1997) Endothelin-1 is not involved in hemoglobin associated vasoactivities. Artif Cells Blood Substit Immobil Biotechnol 25:135-140
29. Zilletti L, Ciuffi M, Franchi-Micheli S, Fusi F, Gentilini G, Moneti G, Valoti M, Sgaragli GP (1994) Cyclooxygenase activiy of hemoglobin. Methods Enzymol 231:562-573

Part 2

NO and Cell Function

Nitro Oxide And Ovarian Cancer

Robin Farias-Eisner, Gautam Chaudhuri [1]

The precise mechanisms by which the host defense to malignant tumors is mediated in vivo is not known. It is now believed that cancer cells themselves may trigger an immune response and this may limit the progression of the tumor. It is also possible to induce non specific resistance to cancer cells therapeutically. Intraperitoneal (i.p.) inoculation of Cornebacterium parvum, which increases nonspecific resistance to murine ovarian teratocarcinoma (MOT) cells can cure mice bearing these tumor cells [1]. Similarly, i.p. administration of Cornebacterium parvum to women [2] was at least partially effective in the treatment of minimal residual ovarian cancer in women. However, the mechanism by which this occurs is not known. Similarly, bacillus Calmette-Guerin (BCG) also appears to be an immunostimulant. When cultured in vitro, peritoneal macrophages obtained from mice previously inoculated with BCG release NO, which is cytostatic or cytolytic for tumor cells [3,4]. This in vitro cytostatic and/or cytolytic activity is also observed after activation of the macrophages by cytokines [5,6]. Interferon γ (INFγ) is important for the priming of macrophages, and tumor necrosis factor α (TNFα) or some other cytokine or bacterial lipopolysaccharide (LPS) is necessary for full induction of activated macrophage cytotoxicity [7,8]. However, the role of NO in mediating tumoricidal activity in vivo and the mechanism by which this occurs is not exactly known. We therefore evaluated the role of NO in mediating both tumoricidal activity in vivo and the mechanism by which this occurs. We therefore undertook to evaluate both the in vivo role of NO in mediating the host resistance to a syngeneic ovarian tumor in BCG-inoculated mice as well as the host resistance to a xenogeneic ovarian tumor graft and the role of IFNα and TNFα in this process [9]. We also evaluated the mechanism by which NO causes this cytotoxicity [10].

---

[1] Department of Obstetrics and Gynecology and Melocular and Medical Pharmacology, UCLA School of Medicine, 10833 Le Conte Avenue, Los Angeles. California 90095-1740, USA

For the in vivo studies [9], we utilized two different ovarian cancer cell lines. The MOT cell line utilized to study the mechanism of BCG-induced host resistance to a syngeneic ovarian tumor, spreads throughout the peritoneal cavity and produces ascites in C3HeB/FeJ mice akin to human ovarian cancer [11,12]. In studies related to xenogeneic ovarian tumor grafts, we utilized a human epithelial ovarian cancer cell line, NIH : OVCAR-3. This cell line grows and develops as a lethal tumor only in athymic nude (immunologically incompetent mice [13]. In these studies, we transplanted these two cancer cell lines i.p. into 6 to 10 C3HeB/FeJ mice weighing 17-28 g. which served as our model system.

We observed significant increase in the number of MOT cells and an increase in body weight accompanied by ascites in control (non-BCG treated) animals and in those treated with subcutaneous (s.c.) BCG. Tumoricidal activity in animals preinoculated with i.p. BCG was evident by a reduction in the number of viable intraperitoneal MOT cells and absence of ascites. By contrast, daily i.p. administration (250 mg/kg) of L-N^{ω}-methylarginine (L-NMA), an inhibitor of NO synthesis, prevented the tumoricidal activity in animals preinoculated with i.p. BCG and subsequently transplanted with MOT cells. In this group we observed that the MOT cells proliferated to the same extent as that of control animals which received MOT cells alone. In separate groups of animals, in addition to receiving L-NMA the animals also received l-arginine or d-arginine. We observed that l-arginine and not d-arginine reversed the effects of L-NMA. These results indicated to us that only i.p. administration of BCG led to tumoricidal activity which was mediated by NO formation. In order to further assess the intermediate role of IFNγ in the BCG induced NO formation thereby leading to tumoricidal activity, IFNγ antibody (IFNγ AB) or TNFα AB were infected (200 μg/animal) i.p. IFNγ antibody (IFNγ AB) but not TNFα AB blocked the tumoricidal activity of i.p. BCG administration. These results indicated to us that BCG acted as an immunostimulant by stimulating NO formation most likely by the macrophages with INFγ playing an intermediate role in this process.

By contrast when OVCAR cells were transplanted i.p. into mice, marked tumoricidal activity was observed in these animals. No ascites was observed nor was there any mortality in these animals. Intraperitoneal administration of L-NMA to OVCAR transplanted mice reduced tumoricidal activity and resulted in ascites formation and l-arginine reversed the effects of L-NMA. Intraperitoneal administration of IFNγ AB and TNFα AB did not affect the tumoricidal activity.

These results indicated to us the importance of NO in mediating both BCG-induced host resistance to a syngeneic ovarian tumor as well as host resistance to xenogeneic ovarian tumor grafts. Our results also demonstrated the importance of

IFNγ but not TNFα as an important mediator of the BCG non-specific host resistance against syngeneic ovarian tumor cells in vivo. Results from these studies have potential significance. Our observations may explain why local instillation of BCG into the bladder is effective in the treatment of superficial muscle bladder cancers in humans [14,15]. Our results may also explain why s.c. and intravenous administration of BCG to women was not effective in the treatment of human ovarian cancer [16], whereas i.p. instillation of C parvum, which like BCG stimulates the host defense mechanism [17], was at least partially effective in the treatment of minimal residual ovarian cancer in women [18].

In further investigations on the mechanism of tumoricidal activity of NO we investigated the mechanism of cytotoxicity of the NO donor 3-morpholino-sydnonimine towards the OVCAR cell line. We observed [10] that the NO mediated loss of cell viability was dependent on both NO and hydrogen peroxide (H_2O_2). Superoxide and its reaction product with NO, peroxynitrite, did not appear to be directly involved in the observed NO-mediated cytotoxicity against the cancer cell line. On this basis, we proposed that NO or activated macrophage mediated cytotoxicity against this cancer cell line can be attributed to the generation of reactive radical species, such as .OH through a chemical process involving a trace redox active metal, N_2O_2 and NO. This hypothesis may also be the basis for explaining the differential susceptibility of cells to NO cytotoxicity. Because NO does not affect the gluathione peroxidase-glutathione reductase system, cells that rely heavily on these enzymes for handling intracellular H_2O_2 would have increased resistance to NO cytotoxicity.

References

1. Knapp RC, Berkowitz RS (1977) Corynebacterium parvum as an immunotherapeutic agent in an ovarian cancer model. Am J Obstet Gynecol 128: 782-786
2. Bast RC Jr, Berek JS, Obrist R, Griffiths CT, Berkowitz RS, Hacker NF, Parker L, Lagasse LD, Knapp RC (1983) Intraperitoneal immunotherapy of human ovarian carcinoma with Corynebacterium parvum. Cancer Res 43: 1395-1401
3. Hibbs JB Jr, Taintor RR, Chapman HA Jr, Weinberg JB (1977) Macrophage tumor killing: influence of the local environment. Science 197: 279-282
4. Cleveland RP, Meltzer MS, Zbar B (1974) Tumor cytotoxicity in vitro by macrophages from mice infected with mycobacterium bovis strain BCG. J Natl Cancer Inst 52: 1887-1895
5. Pace JL, Russell SW, Torres BA, Johnson HM, Gray PW (1983) Recombinant mouse gamma interferon induces the priming step in macrophage activation for tumor cell killing. J Immunol 130: 2011-2013
6. Lorsbach RB, Murphy WJ, Lowenstein CJ, Snyder SH, Russell SW (1993) Expression of the nitric oxide synthase gene in mouse macrophages activated for tumor cell killing. Molecular basis for the synergy between interferon-gamma and lipopolysaccharide. J Biol Chem 268: 1908-1913

7. Drapier JC, Wietzerbin J, Hibbs JB Jr (1988) Interferon-gamma and tumor necrosis factor induce the L-arginine-dependent cytotoxic effector mechanism in murine macrophages. Eur J Immunol 18: 1587-1592
8. Ding AH, Nathan CF, Stuehr DJ (1988) Release of reactive nitrogen intermediates and reactive oxygen intermediates from mouse peritoneal macrophages. Comparison of activating cytokines and evidence for independent production. J Immunol 141: 2407-2412
9. Farias-Eisner R; Sherman MP; Aeberhard E; Chaudhuri G (1994) Nitric oxide is an important mediator for tumoricidal activity in vivo. Proc Natl Acad Sci 91: 9407-9411
10. Farias-Eisner R, Chaudhuri G, Aeberhard E, Fukuto JM (1996) The chemistry and tumoricidal activity of nitric oxide/hydrogen peroxide and the implications to cell resistance/ susceptibility. J Biol Chem 271: 6144-6151
11. Bast RC Jr, Knapp RC, Donahue VC, Thurston JG, Mitchell AK, Feeney M, Schlossman SF (1980) Specificity of heteroantisera developed against purified populations of intact murine ovarian carcinoma cells. J Natl Cancer Inst 64: 365-372
12. Feldman GB, Knapp RC, Order SE, Hellman S (1972) The role of lymphatic obstruction in the formation of ascites in a murine ovarian carcinoma. Cancer Res 32: 1663-1666
13. FitzGerald DJ, Bjorn MJ, Ferris RJ, Winkelhake JL, Frankel AE, Hamilton TC, Ozols RF, Willingham MC, Pastan I (1987) Antitumor activity of an immunotoxin in a nude mouse model of human ovarian cancer. Cancer Res 47: 1407-1410
14. Morales A, Ottenhof P, Emerson L (1981) Treatment of residual, non-infiltrating bladder cancer with bacillus Calmette-Guerin. J Urol 125: 649-651
15. van der Meijden AP, Steerenberg PA, van Hoogstraaten IM, Kerckhaert JA, Schreinemachers LM, Harthoorn-Lasthuizen EJ, Hagenaars AM, de Jong WH, Debruijne FM, Ruitenberg EJ (1989) Immune reactions in patients with superficial bladder cancer after intradermal and intravesical treatment with bacillus Calmette-Guerin. Cancer Immunol Immunother 28: 287-295
16. Alberts DS, Mason-Liddil N, O'Toole RV, Abbott TM, Kronmal R, Hilgers RD, Surwit EA, Eyre HJ, Baker LH (1989) Randomized phase III trial of chemoimmunotherapy in patients with previously untreated stage III, optimal disease ovarian cancer: a Southwest Oncology Group Study. Gynecol Oncol 32: 16-21
17. Likhite VV, Halpern BN (1974) Lasting rejection of mammary adenocarcinoma cell tumors in DBA-2 mice with intratumor injection of killed Corynebacterium parvum. Cancer Res 34: 341-344
18. Bast RC Jr, Berek JS, Obrist R, Griffiths CT, Berkowitz RS, Hacker NF, Parker L, Lagasse LD, Knapp RC (1983) Intraperitoneal immunotherapy of human ovarian carcinoma with Corynebacterium parvum. Cancer Res 43: 1395-1401

Nitric Oxide Production and Long-term Potentiation in the Rat Hippocampus Following Transient Cerebral Ischemia

Hiroko Togashi, Ken-ichi Ueno, Kiyoshi Mori, Machiko Matsumoto, Yoshitada Itoh, Katsura Shinohara, Mitsuhiro Yoshioka* [1]

SUMMARY. The cerebral dysfunction such as deterioration of memory remains serious complications of ischemic reperfusion injury. Long-term potentiation (LTP) has been widely studied as a form of synaptic plasticity that represents a cellular mechanism of learning and memory. Numerous processes and molecules are reported to be involved in LTP mechanisms, and some elements including neurotrophic and transcription factors are likely to be common with those involved in neural death after cerebral ischemia. Nitric oxide (NO) is a molecule which has crucial roles in neuronal injury. In our study, we focused on LTP formation as a functional response to cerebral ischemia and elucidated the implication of NO production in LTP formation in the rat hippocampus following transient cerebral ischemia. NO production was evaluated by oxidative NO metabolite levels determined using *in vivo* brain microdialysis. Transient cerebral ischemia produced a marked inhibition of LTP in both Schaffer-CA1 synapses and perforant path-dentate gyrus synapses. The increase in hippocampal NO production was observed to precede LTP inhibition. Direct or indirect inhibition of an inducible NO synthase (iNOS) rescued ischemia-induced LTP inhibition. Centrally administered bacterial endotoxin, lipopolysaccharide, which is known to induce iNOS expression, could mimic the time-course changes in hippocampal NO production observed after ischemic insult. These findings suggest that iNOS-derived NO is partly responsible for the ischemia-induced impairment of LTP in the rat hippocampus.

KEY WORDS: Cerebral ischemia, Long-term potentiation, Nitric oxide, Hippocampus

---

[1] Department of Pharmacology, Hokkaido University Graduate School of Medicine and *Department of Anesthesiology, Hokkaido University School of Medicine, Sapporo 060-8638, Japan

Introduction

Cerebral ischemia is a major cause of neurodegenerative diseases. Neural death occurs a few days after transient cerebral ischemia. Irreversible cell damage of the neurons results in profound cerebral dysfunction such as memory and motor deficits. The deterioration of cerebral function remains serious complications of ischemic reperfusion injury. Among the cerebral regions, hippocampal and cortical neurons are known to contain vulnerable regions to ischemia and/or oxidative stress during the pathological event [1,2]. In addition, it has been reported that transient ischemia can produce cerebral dysfunction without tissue damage such as loss of neurons and innervating pathways [3]. Recently, we also reported that an incomplete cerebral ischemia produced by bilateral carotid artery occlusion elicited a delayed electrophysiological changes in the rat hippocampus without any histological changes [4]. Thus, it is likely that the cerebral dysfunction accompanied with transient ischemia precedes or is not always associated with histological changes.

Long-term potentiation (LTP) is an electrophysiological phenomenon that brief trains of high- frequency stimulation to the excitatory pathways elicit a long lasting increase in the efficiency of synaptic transmission. LTP was first described by Bliss and Lømo in the perforant path-dentate gyrus (DG) synapses of the anesthetized rabbits [5]. There is growing evidence that it is implicated in certain forms of memory. Thus, LTP has been widely studied as a cellular mechanism of learning and memory in several brain regions (reviewed in ref. 6). However, numerous reports are concerned on *in vitro* studies using brain slices where neuronal circuit connections are not always retained. To understand the consequence underlying pathological events after ischemia-reperfusion, the *in vivo* profile should be explored as well.

Nitric oxide (NO), a gaseous cell signaling molecule, is synthesized via the activation of NO synthase (NOS) . The widespread but topographical localization of NOS isoforms, inducible NOS (iNOS), neuronal NOS (nNOS) and endothelial NOS (eNOS), in neuronal and glial cells [7] supposes diverse roles of NO in the pathophysiology of neurodegenerative diseases such as cerebral ischemia. Concerning LTP formation, NO is thought to exert double-edged effects. Under physiological conditions, LTP is triggered by the Ca^{++} entry through NMDA receptors located on the postsynaptic cells. Thereafter, the potentiated response is maintained in part by presynaptic mechanisms. It is postulated that an intercellular molecule is released as a retrograde messenger from the postsynaptic sites to increase a transmitter release from the presynaptic sites. NO is a candidate for a retrograde messenger (reviewed in ref. 6). On the contrary, NO is expected to deteriorate LTP formation under certain circumstances, since pharmacological blockade of NO synthesis ameliorates neuronal cell death following cerebral ischemia [8]. However, it is not conclusive whether the change in LTP formation that is attributed to cerebral ischemia is resulted from excessive NO production and, if so, which isoforms of NOS are concerned.

We now focus on LTP formation as a functional response to transient cerebral ischemia, in the hippocampus, a critical brain region for learning and memory function. Particular emphasis is placed on the implication of NO production in the development of ischemia-induced LTP impairment.

Transient cerebral ischemia and LTP

LTP is thought to represent a form of synaptic plasticity and has been exclusively studied as a model of the neuronal basis of learning and memory. In the hippocampus, LTP was observed in all excitatory pathways; i.e. the perforant path which is a main pathway originated from entorhinal area-DG granule cell synapses, mossy fiber-CA3/CA4 synapses, Schaffer collaterals projected from fimbria-CA1/CA2 synapses (reviewed in ref. 6). Histochemical studies indicate that the CA1 region is vulnerable to ischemic insult and/or oxidative stress [1,2]. As listed in Table 1, numerous processes and molecules are reported to be involved in the LTP formation mechanisms. Some including neurotrophic and transcription factors have been described as those in neurodegenerative processes after cerebral ischemia (Table 2).

We studied the changes of LTP formation in Schaffer-CA1 and perforant path-DG synapses. Two ischemic rat models which are well-studied preparations as global ischemic models were used; 2-vessel occlusion (2VO) and 4-vessel occlusion (4VO) models with an incomplete and complete cerebral ischemia, respectively [1,36].

Surgical procedure: Wistar male rats (12-15 weeks old) were subjected to electrical permanent occlusion of bilateral vertebral arteries (for 4VO model) or to sham operation (for 2VO model) under halothane anesthesia. Seven days therafter, bilateral common carotid arteries were occluded for 10 minutes by the ligature technique [36].

Recording of LTP: Four days after occlusion of common carotid arteries, LTP was recorded. The stimulating electrode was inserted into the Schaffer collateral region or the perforant path, and the recording electrode was inserted into the pyramidal cell layer of CA1 or the granule cell layer of DG in the hippocampus. Electrical stimulation, which was consisted of 5 or 10 trains at 1 Hz each composed of 8 pulses at 400 Hz, was delivered as tetanic stimulation. The amplitude of the population spike was measured in both synapses and that evoked by pre-tetanic stimulation was defined as 100 %. The area under the curve (AUC) of the time-course changes after tetanic stimulation was also evaluated over 1 hour. Experimental protocols are presented in Fig. 1.

Table 1. Possible elements implicated in hippocampal LTP formation

Elements	Effects on LTP induction	Effects on LTP maitenance	Evaluated by preparation/ manipulation	Ref.
<Receptors >				
NMDA	+	+	rat, slice, CA1/ glutamate	*
	+	no	rat, slice, CA1/ NMDA	*
GluRε1	(+)		mouse, slice, CA1/ mutation	*
Glycine binding sites	+		neonatal rat, slice, CA1/ antagonist	*
AMPA	no	no	rat, slice, CA1/ AMPA	*
	no	+	rat, slice, CA1/ AMPA	*
mGluR	+		rat, conscious, DG/ MCPG	*
	+		rat, anesthetized, DG/ MCPG	*
mGluR1	+		mouse, slice, CA1/ mutation	*
$GABA_B$	+		rat, slice/ antagonist	*
benzodiazepine	-		rat, slice/ diazepam	*
ACh muscarinic	+	+	rat, slice, CA1 / carbachol, atropine	*
nicotinic	(+)		rat, slice, CA1/ agonist	*
dopamine	+	no	conscious rabbit, DG/ haloperidol	*
serotonin $5\text{-}HT_3$	+	+	rat, slice, CA1/ 2-methyl-5-HT, antagonist	*
adenosine A_1	-		rat, slice, CA1/ adenosine, A_1 antagonist	*
opioid μ	(+)	(+)	rat, anesthetized, CA3/ DAMGO	*
κ	-	no	guinea pig, slice, DG/ U69,593, NBNI	*
CCK	no	+	rat, slice, DG/ CCK 8-S	*
glucocorticoid	(+/-)		mouse, slice, CA1/ corticosterone	*
	(+/-)		rat, slice, CA1/ corticosteroid, mepyrapone	*
<protein kinases>				
CaMK I	+		rat, slice, CA1/ antibody	9
CaMK II	+	+	mouse, slice, CA1/ mutation	*
	+		rat(9-13 days old), slice, CA1/ VV infection	*
	+		rat, slice, CA1/ antibody	9
CaMK IV	+		rat, slice, CA1/ antibody	9
PKA	no	+	rat, slice, CA1/ inhibitor	*
	+	+	rat, slice, CA3/ inhibitor	*
	+	+	mouse, slice, DG/ mutation	10
RIβ/Cβ1 subunit	+	+	mouse, slice, CA3/ mutation	*
RIβ subunit		no	mouse, slice, CA1/ mutation	*
PKG	+		rat, slice, CA1/ activator, inhibitor	*
	+		rat, slice, CA1/ inhibitor	*
	+		rat, slice, CA1/ inhibitor(KT5823)	*
PKC	+	+	rat, slice, CA1/ inhibitor	*
	+		rat, slice, CA1/ SOD & catalase, PKC activity	11
γ subtype	(+)		mouse, slice, CA1/ mutation	*
tyrosine kinase Fyn	+	+	mouse, slice, CA1/ mutation	*
	+		mouse, slice, CA1/ transgenic	12
<nucleotides>				
ATP	no	+	rat, slice, CA1/ ATP & ATP analog	*
cyclic AMP	no	+	rat, slice, CA1/ antagonist	*
	+	+	rat, slice, CA3/ antagonist	*
	(+)	(+)	mouse, slice, CA1/ mutation(AC-I)	*
cyclic GMP	+	+	rat, slice, CA1/ 8-Br-cGMP	*
	+		rat, slice, CA1/ GC inhibitor(LY83583)	*

Table 1. (continued)

Elements	Effects on LTP induction/	maitenance	Evaluated by preparation/ manipulation	Ref.
<others>				
Ca^{2+}	+		rat, slice, CA1/ Ca^{2+}	*
K^{+}	+	+	rat, slice, CA1/ K^{+} channel blocker	*
NO	+		rat, slice, CA1/ NOS inhibitor	*
	+		rat, cultured neuron, CA1 / NO scavanger, NOS inhibitor, NO donor	*
	+		rat, anesthetized, CA1/ nNOS inhibitor	*
	+		mouse, slice, CA1/ eNOS&cNOS mutation	*
CO	+		rat, slice, CA1/ heme oxygenase inhibitor	*
PLA_2	+	no	rat, slice, CA1/ inhibitor	13
arachidonic acid	+	+	rat, anesthetized, DG / measure arachidonic acid	*
	+	+	slice, CA1/ arachidonic acid, diacylglycerol	14
PAF	no	+	rat, slice, CA1/PAF antagonist(BN-52021)	15
GAP-43	+		mouse, slice, CA1/ mutation	*
	+		mouse, slice, CA1/ transgenic	*
	+		mouse, anesthetized, DG/ transgenic, GAP-43 mRNA expression	16
calpain	+		rat with genetic deficiency in the endogenous calpain inhibitor	*
	+		rat, slice, CA1/ antibody	17
neurotrophic factors				
IL-1β	(-)		rat, slice, CA1/ IL-1β	*
	-		mouse, slice, CA3/ IL-1β, tripeptide analog	*
	-		rat, in vitro, DG/ IL-1β, IL-1ra	18
IL-2	-	-	rat, slice, CA1/ IL-2	*
IL-6	-		rat, slice, CA1/ rhIL-6, IL-6 antibody	19
BDNF	no	+	neonatal rat, slice, CA1/ BDNF	*
	no	+	mouse, slice,CA1/ mutation	*
	no	+	rat, slice, CA1/ TrkB antiserum & TrkB-IgG fusion protein	20
	no	+	rat, in vivo, DG/ mRNA expression	21
transcription factors				
CREB α&δ	no	+	mouse, slice, CA3/ mutation	*
		+	mouse, slice, CA1, CA3/ mutation (t-PA)	22

Recent references concerned with hippocampal LTP formation are cited. *For detail, refer to the reviews (ref. 6 and ref. 23).

no, no effect; ( ), modulation(facilitation (+), attenuation (-)); (+/-), inversed-U shape modulation

abbreviations: AC, adenylyl cyclase; ACh, acetylcholine; AMPA, α-amino-3-hydroxy-5-metghyl-4-isoxazole-propionic acid; BDNF, brain-derived neurotrophic factor; CREB, cyclic AMP response element binding protein; CaMKII, calcium/calmodulin-dependent kinase II; CCK, cholecystokinin; DG, dentate gyrus; GABA, γ-aminobutylic acid; GC, guanylyl cyclase; IL, interleukin; MCPG, (R,S)-α-methyl-4-carboxyphenylglycine, mGluR, metabotropic glutamate receptor; NBNI, norbinaltorphimine; NMDA, N-methyl-D-aspartate; PAF, platelet-activating factor; PKA, cyclic AMP-dependent protein kinase, A kinase; PKC, calcium-phospholipid-dependent protein kinase, C kinase; PKG, cyclic GMP-dependent protein kinase, G kinase; PLA_2, phospholipase A_2; VV, vaccinia virus

Table 2. Possible neurotrophic factors related to cerebral ischemic injury

<neurotrophic factors>	<abbreviation>	Ref.
neurotrophins		
nerve growth factor	NGF	24
brain-derived neurotrophic factor	BDNF	25
neurotrophin-3	NT-3	24
neurotrophin-4	NT-4	26
neurotrophin-5	NT-5	26
FGF family		
basic fibroblast growth factor	bFGF	27
acidic fibroblast growth factor	aFGF	28
insulin like growth factors (I / II)	IGF	29
epidermal growth factor	EGF	27
vascular endothelial growth factor	VEGF	30
platelet derived growth factor	PDGF	31
hepatocyte growth factor	HGF	32
TGF superfamily		
transforming growth factor (β_1 / β_2)	TGF	33
grial cell-derived neurotrophic factor	GDNF	34
ciliary neurotrophic factor	CNTF	35

modified from ref. 6 and ref. 23. Recent references concerned with hippocampal LTP formation are cited.

LTP formation was attenuated by 2VO and 4VO in both Schaffer collateral-CA1 and the perforant path-DG synapses. Although the CA1 region is thought to be the most vulnerable to ischemia as compared to other cerebral regions [1,2], ischemia-induced LTP inhibition was observed at the same degree in both hippocampal synapses in our studies. In addition, the response of LTP formation in 2VO model was almost comparable to that in 4VO model. Thus, it is unlikely that the degree of LTP inhibition associated with transient ischemia is that expected

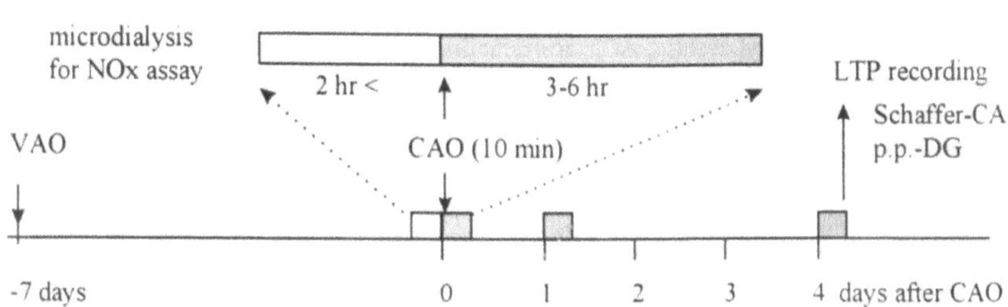

Fig. 1. Schematic drawings for experimental protocols. CAO, carotid artery occlusion; DG, dentate dyrus; p.p., perforant path; VAO, vertebral artery occlusion.

from the vulnerability to ischemia. However, we have found that some neurotropic agents significantly ameliorated LTP impairment in the DG synapses but not in the CA1 synapses [37,38]. For instance, we found that pretreatment with interleukin (IL)-1β antagonist restored LTP in both synapses in 2VO model. On the contrary, in 4VO model, LTP was restored in the DG, not in the CA1 region. Our pharmacological evidence may reflect the different degree of LTP impairment between the synapses, the vulnerability of CA1 synapses to cerebral ischemia. We have, recently, reported that 2VO did not produce histological changes such as loss of neurons and innervating pathways [4]. In other words, these findings indicate that LTP recording is a sensitive method for detecting functional changes implicated in cerebral ischemia. AUC of the time-course changes in LTP is shown in Table 3.

Table 3. The changes in LTP formation after transient cerebral ischemia

	Control	2VO	4VO
(n)	(8)	(9)	(9)
CA1	10078±659	6863±537*	7016±399*
DG	11063±746	6957±697*	6024±241*

The value indicates AUC (%·min) of the LTP formation curves determined over 2 hours after tetanic stimulation. LTP was recorded 4 days after ischemic insults (10 min). *$p<0.05$ vs. Control.

Transient cerebral ischemia and hippocampal NO production

NO is a molecule which is thought to have a dual action on LTP formation (reviewed in ref. 8). Under physiological conditions, it is postulated that NO is released from the postsynaptic sites, and acts on the presynaptic sites as a retrograde molecule, where enhances an excitatory neurotransmitter release and results in long-lasting increase in synaptic efficacy. Pharmacological blockade of NO synthesis has been observed to suppress the LTP formation. We also reported that a broad-spectrum NOS inhibitor, N^{ω}-nitro-L-arginine methyl ester (L-NAME), inhibited LTP formation in DG synapses of the anesthetized rat [39]. Co-injection of L-arginine reversed the effect of L-NAME. On the contrary, much evidence is accumulated that NO might be involved in the pathogenesis of cerebral ischemia-induced dysfunction such as LTP inhibition. However, there is little direct evidence, implying that NO is released into the extracellular space.

We now intended to describe the dynamic changes in NO production following transient cerebral ischemia in the rat hippocampus and to clarify whether ischemia-induced impairment of LTP formation is implicated in excessive NO production. We determined nitrite (NO_2^-) and nitrate (NO_3^-) as an index of NO production using an *in vivo* brain microdialysis [40,41].

Measurement of NO metabolites: Two days before common carotid artery occlusion, a guide cannula was implanted into the rat hippocampus under ketamine anesthesia. A microdialysis probe was perfused with Ringer's solution at a rate of 1 μl/min. The dialysate was collected every 10 min and was injected to a NO detector-HPLC system (ENO-10; Eicom, Kyoto, Japan) via on-line auto injector. The concentration of NO_2^- and NO_3^- in the dialysate was measured by Griess method. NOx (NO_2^- plus NO_3^-) levels before clamping common carotid arteries were defined as 100 %, and were determined over 3-6 hr and on 1 and 4 days after occlusion.

2VO and 4 VO produced a transient increase in hippocampal NOx levels. The peak responses were significantly different from those in the sham-operated rat and dependent upon the degree of ischemic insults. Moreover, 4VO produced a gradual increase in NOx levels 3 hr after reperfusion. The NO metabolite levels measured 1 and 4 days after 4VO were higher than those in the sham-operated rats (Table 4). Of note is the fact that the changes in NOx levels preceded hippocampal LTP inhibition, since our previous study showed that LTP inhibition was more potent 4 days than 1 day after 2VO [4]. An inducible NO synthase inhibitor, aminoguanidine, significantly abolished the increase in NOx levels. In addition, intracerebroventricular infusion of endotoxin, lipopolysaccharide (LPS), which is known to induce iNOS expression [42], produced a marked increase in hippocampal NOx levels (Fig. 2). The time-course changes after LPS treatment was similar to those observed after ischemic insult. These findings suggest that the changes in the hippocampal NOx levels might reflect NO production in the post-ischemic brain, which was derived from iNOS. On the other hand, in 2VO model, both 1 day and 4 days after occlusion, NOx levels did not show any significant changes. It is still unclear whether the difference in the NOx dynamic profiles observed between 2VO and 4VO models is essential or merely due to methodological limitation for measuring oxidative NO metabolites as an index of NO production.

Discussion

In the pathogenesis of cerebral ischemia, it is postulated that the roles of NO might be dependent upon the stage of evolution of the ischemic process [8]. Cerebral ischemia is known to produce an increase in protein expression of iNOS as well as endothelial and neuronal constitutive NOS. However, it is not fully understood whether NO production following ischemia is implicated in cerebral dysfunction and, if so, which NOS isoforms are responsible for it. In these view points, it is very important to determine the time-course changes in NO production in the post-ischemic brain. We found that 4VO produced a biphasic increase in oxidative NO metabolites. An iNOS inhibitor aminoguanidine abolished the response observed

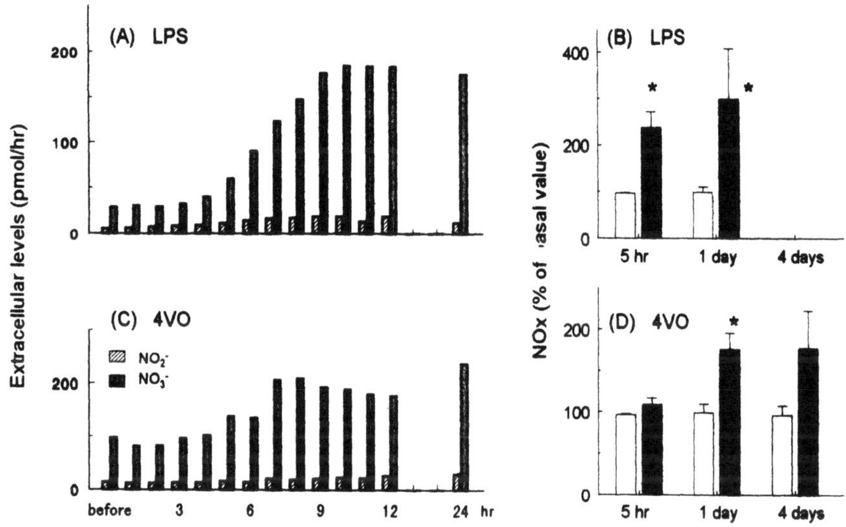

Fig. 2. Hippocampal NO metabolite levels after LPS administration and 4-vessel occlusion. (A)&(C) representatives showing the time-course changes of NO_2^- and NO_3^- levels. (B)&(D) the changes in NOx levels 5 hrs, 1 and 4 days after LPS administration (0.1 mg/kg, i.p.) and 4-vessel occlusion (4VO, 10 min), respectively. Values indicate mean ± SEM in control rats (open column, n=6) and test rats (closed columns, n=6-8). $*p<0.05$ v.s. corresponding controls.

24 hours after reperfusion, without affecting the response observed at an early stage [41]. Thus, iNOS-mediated mechanism is likely to contribute to NO production following transient cerebral ischemia at a late stage. Indeed, iNOS mRNA is known to be induced 6-12 hours after cerebral ischemia [8,43].

It has been reported that certain cytokines, such as tumor necrosis factor-α (TNF-α) and IL-1β, are released in the central nervous system following cerebral ischemia [44]. The previous studies have shown that the expression of IL-1β, a proinflammatory cytokine, was upregulated after focal brain ischemia [45,46]. IL-1β receptor antagonist attenuated neuronal cell death [47]. We have recently reported IL-1β receptor antagonist ameliorated 2VO- and 4VO-induced LTP impairment [38]. Thus, IL-1β is likely to be involved in the mechanisms underlying LTP inhibition following transient ischemia. LPS is reported to induce iNOS mRNA in the rat brain, accompanied with a preceded and marked rise in IL-1β mRNA [42]. We also found that LPS produced a marked increase in oxidative NO metabolite levels in the rat hippocampus as well. These findings further support a detrimental role of iNOS-derived NO in the post-ischemic brain. This would in turn suggest that the blockade of iNOS/NO pathway might provide new therapeutic strategies for reperfusion injury. Numerous neurotrophic and transcription factors are also proposed to have crucial roles in ischemia-induced neuronal injury (Table 4), however, their implication in cerebral dysfunction such as LTP inhibition or NO production is far from conclusion.

It has recently been reported that nNOS-derived NO may play a pivotal role in the brain damage observed in the post-ischemic brain. A selective nNOS inhibitor elicited a dose-dependent reduction of infarction volume in rats with transient middle cerebral artery (MCA) occlusion [48 ,49]. Neuronal NOS is known to be inducible under certain circumstances. It might be possible that the two NOS isoforms act critical roles in the post-ischemic brain. However, it remains unsolved whether a nNOS inhibitor would ameliorate LTP inhibition following cerebral ischemia.

We have studied on the changes of LTP formation, as a functional response to transient cerebral ischemia. LTP inhibition was observed in an incomplete ischemia, where microscopic tissue damage was not evident. Thus, LTP is a sensitive index to evaluate a functional response to ischemic insult and to understand the mechanisms involved in cerebral dysfunction accompanied with transient cerebral ischemia. This electrophysiological approach might become more important to provide insights into therapeutic strategies for reperfusion injury and to delay or prevent multiple degenerative diseases.

Table 4. The changes in hippocampal NOx levels after transient cerebral ischemia

	control	2VO	4VO
(n)	(8)	(9)	(8)
early phase	99.5± 2.2	114.3±14.7	133.2±10.0*
1day	99.5±10.7	103.3±26.1	175.8±20.9*
4days	96.4±12.2	111.1±34.1	177.1±48.2

The value indicates % of NOx (NO_2^- plus NO_3^-) levels as the basal level obtained before common carotid artery occlusion was defined as 100 %. *$p<0.05$ vs control.

References

1. Pulsinelli WA, Brierley JB (1979) A new model of bilateral hemispheric ischemia in the unanesthetized rat. Stroke 10: 267-272
2. Kirino T, Sano K (1984) Fine structural nature of delayed neuronal death following ischemia in the gerbil hippocampus. Acta Neuropathol (Berl) 62: 209-218
3. Astrup JA, Siesjö Bo K, Symon L (1981) Thresholds in cerebral ischemia - the ischemic penumbra. Stroke 12: 723-725

4. Mori K, Yoshioka M, Suda N, Togashi H, Matsumoto M, Ueno K, Saito H (1998) An incomplete cerebral ischemia produced a delayed dysfunction in the rat hippocampal system. Brain Res: 795: 221-226
5. Bliss TVP, Lømo T (1973) Long-lasting potentiation of synaptic transmission in the dentate area of the anesthetized rabbit following stimulation of the perforant path. J Physiol 232: 331-356
6. McEachern JC, Shaw CA (1996) An alternative to the LTP orthodoxy: a plasticity-pathology continuum model. Brain Res Rev 22: 51-92
7. Nathan C, Xie QW (1994) Nitric oxide synthases: roles, tolls, and controls (review). Cell 78: 915-918
8. Iadecola C (1997) Bright and dark sides of nitric oxide in ischemic brain injury. TINS 20: 132-139
9. Tokuda M, Ahmed BY, Lu YF, Matsui H, Miyamoto O, Yamaguchi F, Konishi R, Hatase O (1997) Involvement of calmodulin-dependent protein kinases-I and –IV in long-term potentiation. Brain Res 755: 162-166
10. Villacres EC, Wong ST, Chavkin C, Storm DR (1998) Type I adenylyl cyclase mutant mice have impaired mossy fiber long-term potentiation. J Neurosci 18: 3186-3194
11. Klann E, Roberson ED, Knapp LT, Sweatt JD (1998) A role for superoxide in protein kinase C activation and induction of long-term potentiation. J Biol Chem 273: 4516-4522
12. Kojima N, Wang J, Mansuy IM, Grant SG, Mayford M, Kandel ER (1997) Rescuing impairment of long-term potentiation in fyn-deficient mice by introducing Fyn transgene. Proc Natl Acad Sci USA 94: 4761-4765
13. Massicotte G, Oliver MW, Lynch G, Baudry M (1990) Effects of bromophenacyl bromide, a phospholipase A2 inhibitor, on the induction and maintenance of LTP in hippocampal slices. Brain Res 537: 49-53
14. Bramham CR, Alkon DL, Lester DS (1994) Arachidonic acid and diacylglycerol act synergistically through protein kinase C to persistently enhance synaptic transmission in the hippocampus. Neuroscience 60, 737-743
15. Kato K, Clark GD, Bazan NG, Zorumski CF (1994) Platelet-activating factor as a potential retrograde messenger in CA1 hippocampal long-term potentiation. Nature 367: 175-179
16. Namgung U, Matsuyama S, Routtenberg A (1997) Long-term potentiation activates the GAP-43 promotor: selective participation of hippocampal mossy cells. Proc Natl Acad Sci USA 94: 11675-11680
17. Musleh W, Bi X, Tocco G, Yaghoubi S, Baudry M (1997) Glycine-induced long-term potentiation is associated with structural and functional modifications of alpha-amino-3-hydroxyl-5-methyl-4-isoxazolepropionic acid receptors. Proc Natl Acad Sci USA 94: 9451-9456
18. Coogan A, O'Connor JJ (1997) Inhibition of NMDA receptor-mediated synaptic transmission in the rat dentate gyrus in vitro by IL-1 beta. Neuroreport 8: 2107-2110
19. Li AJ, Katafuchi T, Oda S, Hori T, Oomura Y (1997) Interleukin-6 inhibits long-term potentiation in rat hippocampal slices. Brain Res 748: 30-38
20. Kang H, Welcher AA, Shelton D, Schuman EM (1997) Neurotrophins and time: different roles for TrkB signaling in hippocampal long-term potentiation. Neuron 19: 653-664

21. Morimoto K, Sato K, Sato S, Yamada N, Hayabara T (1998) Time-dependent changes in neurotrophic factor mRNA expression after kindling and long-term potentiation. Brain Res Bull 45: 599-605
22. Huang YY, Bach ME, Lipp HP, Zhuo M, Wolfer DP, Hawkins RD, Schoonjans L, Kandel ER, Godfraind JM, Mulligan R, Collen D, Carmeliet P (1996) Mice lacking the gene encoding tissue-type plasminogen activator show a selective interference with late-phase long-term potentiation in both Schaffer collateral and mossy fiber pathways. Proc Natl Acad Sci USA 93: 8699-8704
23. Togashi H, Yoshioka M (1998) Transient cerebral ischemia and long-term potentiation in the rat hippocampus. Folia Pharmacol Jpn 111: 55-63 (in Japanese)
24. Kokaia Z, Zhao Q, Kokaia M, Elmer E, Metsis M, Smith ML, Siesjo BK, Lindvall O (1995) Regulation of brain-derived neurotrophic factor gene expression after transient middle cerebral artery occlusion with and without brain damage. Exp Neurol 136: 73-78
25. Ferrer I, Ballabriga J, Marti E, Perez E, Alberch J, Arenas E (1998) BDNF up-regulates TrkB protein and prevents the death of CA1 neurons following transient forebrain ischemia. Brain Pathol 8: 253-261
26. Chan KM, Lam DT, Pong K, Widmer HR, Hefti F (1996) Neurotrophin-4/5 treatment reduces infarct size in rats with middle cerebral artery occlusion. Neurochem Res 21: 763-767
27. Hicks D, Heidinger V, Mohand-Said S, Sahel J, Dreyfus H (1998): Growth factors and gangliosides as neuroprotective agents in exitotoxicity and ischemia. Gen Pharmacol 30: 265-273
28. Cuevas P, Carceller F, Reimers D. Saenz de Tejada I, Gimenez-Gallego G (1998) Acidic fibroblast growth factor rescues gerbil hippocampal neurons from ischemic apoptotic death. Neurol Res 20(3): 271-274
29. Beilharz EJ, Bassett NS, Sirimanne ES, Williams CE, Gluckman PD (1995) Insulin-like growth factor II is induced during wound repair following hypoxic-ischemic injury in the developing rat brain. Brain Res Mol Brain Res 29: 81-91
30. Cobbs CS, Chen J, Greenberg DA, Graham SH (1998) Vascular endothelial growth factor expression in transient focal cerebral ischemia in the rat. Neurosci Lett 249: 79-82
31. Sakata M, Yanamoto H, Hashimoto N, Iihara K, Tsukahara T, Taniguchi T, Kikuchi H (1998) Induction of infarct tolerance by platelet-derived growth factor against reversible focal ischemia. Brain Res 784: 250-255
32. Miyazawa T, Matsumoto K, Ohmichi H, Katoh H, Yamashita T, Nakamura T (1998) Protection of hippocampal neurons from ischemia-induced delayed neuronal death by hepatocyte growth factor: a novel neurotrophic factor. J Cereb Blood Flow Metab 18: 345-348
33. Vivien D, Bernaudin M, Buisson A, Divoux D, MacKenzie ET, Nouvelot A (1998) Evidence of type I and type II transforming growth factor-beta receptors in central nervous tissues: changes induced by focal cerebral ischemia. J Neurochem 70: 2296-2304
34. Kitagawa H, Hayashi T, Mitsumoto Y, Koga N, Itoyama Y, Abe K (1998): Reduction of ischemic brain injury by topical application of glial cell line-derived neurotrophic factor after permanent middle cerebral artery occlusion in rats. Stroke 29: 1417-1422

35. Lin TN, Wang PY, Chi SI, Kuo JS (1998) Differential regulation of ciliary neurotrophic factor (CNTF) and CNTF receptor alpha (CNTFR alpha) expression following focal cerebral ischemia. Brain Res Mol Brain Res 55: 71-80
36. Himori N, Matsuura A (1989) A simple technique for occlusion and reperfusion of coronary artery in conscious rats. Am J Physiol 256 (Heart Circ Physiol 25): H1719-H1725
37. Mori K, Yoshioka M, Suda N, Ueno K, Togashi H, Saito H (1997) Effects of bifemelane on imcomplete cerebral ischemia-induced inhibition of LTP in the rat hippocampal neurons in vivo. Jpn J Pharmacol 73 (suppl I), 271P
38. Mori K, Togashi H, Itoh Y, Matsumoto M, Ueno K, Yoshioka M (1998) Effects of transient cerebral ischemia on nitric oxide metabolites and long-term potentiation in vivo: its blockade by IL-1β analog. In: Moncada S Toda N, Maeda H, Higgs EA (ed) The Biology of Nitric Oxide, Part 6. Portland Press, London, pp124
39. Iga Y, Yoshioka M, Togashi H, Saito H (1993) Inhibitory action of N^{ω}-nitro-L-arginine methyl ester on in vivo long-term potentiation in the rat dentate gyrus. Eur J Pharmacol 238: 395-398
40. Shintani F, Kanba S, Nakaki T, Sato K, Yagi G, Kato R, Asai M (1994) Measurement by in vivo brain microdialysis of nitric oxide release in the rat cerebellum. J Psychiatr Neurosci 19: 217-221
41. Togashi H, Mori K, Ueno K, Matsumoto M, Suda N, Saito H, Yoshioka M (1998) Consecutive evaluation of nitric oxide production after transient cerebral ischemia in the rat hippocampus using in vivo brain microdialysis. Neurosci Lett 240: 53-57
42. Jacobs RA, Satta MA, Dahia PLM, Chew SL, Grossman AB (1997) Induction of nitric oxide synthase and interleukin-1β, but not heme oxygenase, messenger RNA in rat brain following peripheral administration of endotoxin. Mol Brain Res 49: 238-246
43. Iadecola C, Zhang F, Casey R, Clark HB, Ross ME (1996) Inducible nitric oxide synthase gene expression in vascular cells after transient focal cerebral ischemia. Stroke 27: 1373-1380
44. Dinarello CA (1988) Biology of interleukin 1. FASEB J 2: 108-115
45. Hagan P, Barks JD, Yabut M, Davidson BL, Roessler B, Silverstein FS (1996) Adenovirus-mediated over-expression of interleukin-1 receptor antagonist reduces susceptibility to excitotoxic brain injury in perinatal rats. Neurosci 75 : 1033-1045
46. Wang X, Barone FC, Aiyar NV, Feuerstein GZ (1997) Interleukin-1 receptor and receptor antagonist gene expression after focal stroke in rats. Stroke 28: 155-161
47. Rothwell NJ, Hopkins SJ (1995) Cytokines and the nervous system II: actions and mechanisms of action. TINS 18: 130-136
48. Yoshida T, Limmroth V, Irikura K, Moskowitz MA (1994) The NOS inhibitor, 7-nitroindazole, decreases focal infarct volume but not the response to topical acetylcholine in pial vessels. J Cereb Blood Flow Metab 14: 924-929
49. Zhang ZG, Reif D, Macdonald J, Tang WX, Kamp DK, Gentile RJ, Shakespeare WC, Murray RJ, Chopp M (1996) ARL 17477, a potent and selective neuronal NOS inhibitor decreases infarct volume after transient middle cerebral artery occlusion in the rats. J Cereb Blood Flow Metab 16: 599-604

Estrogen and Nitric Oxide-Possible Mechanism of Anti-atherosclerotic Effect of Estrogen via Isoforms of Nitric Oxide Synthase

Toshio Hayashi, Kazuyoshi Yamada, Teiji Esaki, Emiko Mutoh, Izumi Itoh, Htsuyo Kano, Navin Kumar Thakur, Yukako Asai, Akihisa Iguchi[1]

SUMMARY. Although the atheroprotective effect of estrogen is well known, the mechanism is not completely understood. We investigated the role of nitric oxide in the atheroprotective effects of estrogen. The basal and stimulated NO response of arteries from rabbits were studied using N^{ω}-monomethyl-L-arginine (L-NMMA), acetylcholine (ACh), and other compounds. The aorta of female rabbits released a greater amount of basal NO than did that of oophorectomized females and male rabbits. The greater basal release of NO in female rabbits was decreased in atherosclerotic vessels in animals fed a high cholesterol diet. We investigated the effect of estrogen on endothelial, neuronal and inducible nitric oxide synthase, (eNOS, nNOS, and iNOS) respectively. Preincubation with a physiological concentration of 17β-estradiol (10^{-12} to 10^{-8} M) over 8 hours significantly enhanced activity and protein concentration of eNOS in endothelial cells of cultured human umblical vein (HUVEC) and of bovine aortas (BAEC). 17β-estradiol also enhanced the release of NO from endothelial cells as measured by NO_2^-/NO_3^-, metabolites of NO. These effects were identified via a estrogen receptor-mediated system. 17β-estradiol (10^{-10} to 10^{-8} M) also enhanced the activity of partially purified nNOS in the cytosolic fraction of rabbit cerebellums by DEAE column-chromatography. Estrogen enhanced the fluorescence of dansyl-calmodulin. This may explain one mechanism of short term action of estrogen. The effect of estrogen on iNOS was contrast to those on eNOS or nNOS. When J774 cells, a murine macrophage cell line, were incubated with

---

[1] Department of Geriatrics, Nagoya University School of Medicine, 65 Tsumura-cho, Showa-ku, Nagoya, 466-8560 Japan

interferon-γ and lipopolysaccharide, iNOS was induced and a large amount of NO was released. Pre- or co-incubation of 17β-estradiol inhibited the induction of iNOS protein and NO release by a receptor mediated system. These results may offer one mechanism for the anti-atherosclerotic effect of 17β-estradiol.

KEY WORDS: nitric oxide, nitric oxide synthase, estrogen, atherosclerosis

Introduction

Estrogen is well-known to retard the development of atherosclerosis. Estrogen replacement therapy suppresses the incidence of cardiovascular disease in postmenopausal women, and reduces the plasma level of LDL-cholesterol while increasing that of HDL-cholesterol [1]. However, this change in lipid profile reportedly accounts for only 50% of the effect of estrogen [2]. Although estrogen is supposed to exert a direct action on the vessel wall, its mechanism of action is incompletely understood. Nitric Oxide (NO) protects against the development of atherosclerosis by producing vascular dilatation and inhibiting monocyte adhesion to endothelium [3]. We hypothesized that the mechanism by which 17β-estradiol inhibits the development of atherosclerosis in female mammals is the increased formation of NO or decrease of degradation of NO in endothelial cells. NO formation may modulate the early events in the development of atherosclerosis. To test this hypothesis, we investigated NO release and its actions on aortic rings obtained from male, female, and oophorctomized female rabbits. We then investigated those NO actions on aortic rings obtained from male and female rabbits fed a high-cholesterol diet (HCD). Because estrogen may act via an NO-mediated system, we investigated the interaction between estrogen and each isoforms of NOS. The effects of estrogen on endothelial, neuronal and inducible nitric oxide synthase (eNOS, nNOS and iNOS, respectively) were investigated respectively in several kinds of cells. These information could propose the NO-mediated action as the one of the anti-atherosclerotic effect of 17β-estradiol.

Materials and Methods

Animals

A total of 60 female and 8 male New Zealand white rabbits weighing 2.5 to 3 kg were obtained from Kitayama Rabis (Ina, Nagano, Japan). All animals were initially fed standard rabbit chow (Oriental Koubo Co. Ltd., Tokyo, Japan) for 2 weeks. In experiment I, 16 female rabbits were then randomly divided into 2 gourps. Rabbits in the second group were oophorectomized. Group (Gp) A was normal females, GpB was oophorectomized females, and GpC was male rabbits. In experiment II, feeding a 2 % cholesterol diet to both the male and female rabbits were done for 10 weeks (GpIII and GpIV) and 15 weeks (GpV and GpVI). As counterpart, male and female rabbits were fed with regular diet for 15 weeks (GpI and GpII). The rabbits were housed individually in stainless-steel cages at 20±3 °C with a 12-hour light cycle, and with free access to water. Feeding was restricted to 100 g per day. Blood samples were taken 24 h after the last feeding. The general appearance of the rabbits was observed daily. This protocol was conducted in conformity with the "GUIDING PRINCIPLES FOR RESEARCH INVOLVING ANIMALS AND HUMAN BEINGS" as described in the American Journal of Physiology.

Assay for lipids

Total cholesterol and triglyceride levels were measured by enzymatic techniques as described previously [4]. High density lipoprotein (HDL)-cholesterol was initially measured after precipitation with phosphotungstate-$MgCl_2$ [5].

Preparation for isometric tension measurement

The rabbit aortic ring preparation was similar to that described by Furchgott and Zawadzki [6]. Briefly, after treatment, rabbits were sacrificed by exanguination after being anesthetized with pentobarbital (50 mg/kg i.v.). The thoracic aortae were carefully removed to protect their endothelial linings, cleared of adhering fat and connective tissue, and cut into transverse rings 3-mm wide. The rings were mounted under 1 g of resting tension on stainless-steel hooks in chambers of 20 ml-capacity and bathed in Kreb's Henseleit solution, (pH 7.4 at 37 °C). Tension was measured isometrically using force displacement (FT 03C) transducers and displayed on a Grass polygraph (Model T04, Nihon Kohden, Tokyo, Japan). In order to elucidate the tone-related release of NO from endothelium-intact aortic rings, moderate vascular tone (35-50 % of the contractile response to 122mM KCl) was induced with low concentrations of the α-adrenergic agonist. phenylephrine [7]. In one series of experiments, concentration dependent contractile responses to NO synthase inhibtor, L-NMMA (1-100 μM) were assessed [7]. Experiments were also conducted to determine the responsiveness of endothelium-intact aortic rings to an endothelium-dependent vasodilator, ACh. and calcium ionophore A-23187. The responsiveness of endothelium-denuded aortic rings to the endothelium-independent vasodilator nitroglycerine (NTG) was also measured. In these experiments, submaximal tension was initially induced with phenylephrine. Cumulative concentration responses were obtained with ACh, A-23187, and NG. To rule out the contribution of prostanoids, indomethacin (5 μM) was added to chambers in some experiments and allowed to incubate for 60 min before the start of the experiments.

Cultured cells

HUVEC were taken from the primary culture using collagenase. BAEC were taken from the thoracic aorta of a bovine fetus.

NOS activity assay

Enzymatic reactions were conducted at 37°C for 15 min in 50mM Tris HCl buffer (pH 7.4) containing 50 mM ^{3}H-L-arginine, 100 μM NADPH, 2 mM $CaCl_2$, 100 μM calmodulin, 10 μM tetrahydrobiopterin, and the other test agents indicated before, in a final incubation volume of 100 μl. The L-[2,3,4,5-^{3}H] arginine HCl was previously purified by anionic exchange chromatography on columns of Dowex AG 1-X8, the OH-form [8]. BAEC and HUVEC were used as the source of eNOS, and rabbit cerebellum was used as the source of nNOS. To investigate the function of calmodulin as a co-factor of constitutive NOS such as eNOS or nNOS, fluorescence of dansyl-calmodulin was studied. The detail of method was previously mentioned [9].

Measurement of NO by NO electrode

The principle of a newly developed NO meter (Intermedic Co., Ltd. Nagoya, Japan) is the measurement of the pA-order redox current between a working electrode and a counter electrode [10].

Measurement of NO_2^-/NO_3^-

The concentration of the nitrite and nitrate (NO_2^-/NO_3^-), metabolites of NO, in the culture medium were determined with an autoanalyzer (TCI-NOx 1000, Tokyo Kasei Kogyo Co., Tokyo, Japan) [11].

Statistical Analysis

Data are expressed as mean±S.E.M. Statistically significance was assessed by student's *t*-test. When more than 3 values were anlalyzed, analysis of variance (ANOVA) was used. If a significant F value was found, Scheffe's test was used to identify differences among groups. A P value less than 0.05 was considered statistically significant.

Results

Gender difference of tone-related basal NO release in rabbits

The concentrations of serum lipids did not differ between male and female rabbits (data not shown). Plasma estrone (E1) and estradiol (E2) concentrations were higher in female rabbits (GpA) when compared with males (GpC) or oophorectomized females(GpB). In studies designed to assess the tone-related release of NO from phenylephrine-precontraced aortic rings, SOD produced concentration-related relaxant responses. The magnitude of the relaxation responses was greater in endothelium-intact aortic rings from GpA rabbits than GpB or GpC rabbits (Fig. 1). Addition of catalase produced no additional changes, and addition of 1 mM oxyhemoglobin, a known inactivator of NO, caused contraction of phenylephrine-precontracted artery (data not shown). In separate experiments, after moderate tone was induced with phenylephrine, both L-NMMA and L-N^G-nitroarginine methylester (L-NAME) elicited contraction in a concentration-dependent manner. The magnitude of contraction produced by L-NMMA was greater in aortic rings from GpA than in those obtained from GpC. ACh and A23187 produced concentration dependent relaxation of endothelium-intact aortic rings and NTG relaxed aortic rings which were denuded of endothelium. No significant differences among relaxant responses were seen in aortic rings from any three groups (GpA, B, and C) of rabbits.

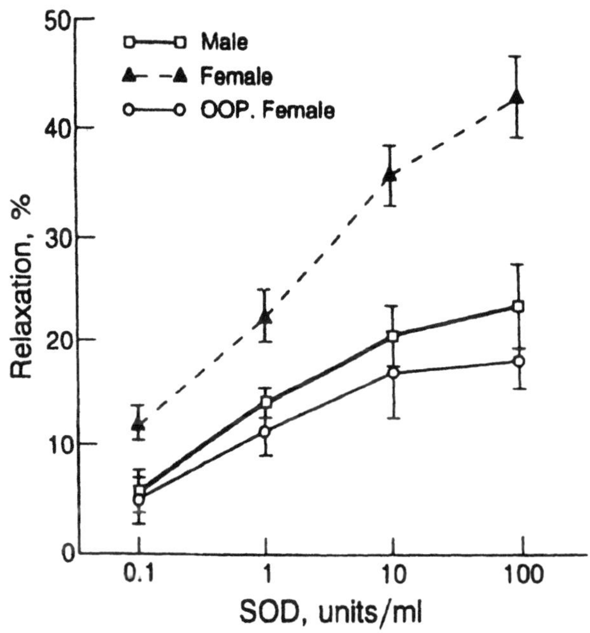

Fig. 1. Percent relaxation of endothelium intact aortic rings to different concentrations of SOD. The aortic rings were moderately contracted, as described in Methods section, before obtaining responses to SOD. The magnitude of relaxation was significantly greater in aortic rings from female rabbits than in aortic rings from either male or oophorectomized (OOP) female rabbits.

Gender difference of tone-related basal NO release in cholesterol-fed rabbits

Feeding a 2 % cholesterol diet to both the male and female rabbits for periods of 10 weeks (GpIII and GpIV) and 15 weeks (GpV and GpVI) raised the concentration of serum lipids in the animals. The serum lipid concentrations did not differ between the sexes and did not differ in those groups fed a 2 % cholesterol diet for 10 weeks (Gp III and Gp IV) as compared with animals fed the HCD for 15 weeks (Gp V and Gp VI)(Table 1). The plasma E1 and E2 concentrations were higher in the females (Gp II) as compared with males in the control group (Gp I). After they had been fed the HCD for 10 and 15 weeks, E1 and E2 levels in these hyperlipidemic male and female rabbits did not differ. There was significantly less aortic atherosclerosis in GpIV than GpIII rabbits fed the HCD for 10 weeks (Fig. 2). In studies designed to assess the tone-related release of NO from aortic rings indirectly, SOD produced concentration-related relaxant responses and the magnitude of these relaxation responses was greatest in endothelium-intact aortic rings obtained from control female rabbits (GpII). The magnitude of relaxant responses from control male (GpI) was significantly less when compared with the control females (GpII). The aortic rings from male and female hypercholesterolemic rabbits relaxed only slightly to SOD, and the magnitude of relaxation was significantly less than that of the controls (GpI and GpII).

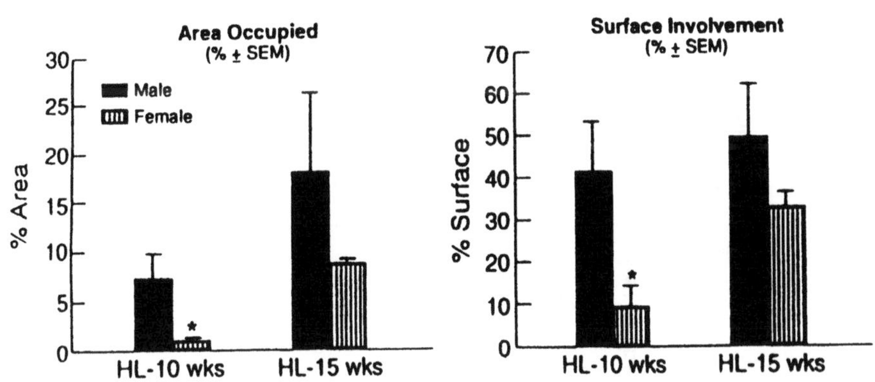

Fig. 2. Gender differences in hyperlipidemic rabbits. Percentage of area occupied by the atherosclerotic lesion as well as the percentage of surface involvement in control animals and animals fed a high cholesterol diet (HCD) for 10 or 15 weekks. The 10 week female HCD animals showed significantly less atherosclerotic involvement of thoracic aorta (*$p<0.05$) as compared to with the corresponding male HCD animals.

Unlike the gender difference to SOD-induced relaxation observed in control animals, no gender difference was observed in hyperchoelsterolemic animals (Fig. 3). In tissues obtained from the various groups, ACh produced concentration-dependent relaxation of PE-precontracted aortic rings that had intact endothelium. In hypercholesterolemic animals, the magnitude of relaxation was diminished, and magnitude was directly related to the duration of hypercholesterolemia (Fig. 4). No significant sex-related differences were noted between the relaxation responses of aortic rings to Ach and A23187 obtained from either conrol or hypercholesterolemic animals. NTG produced concentration-dependent relaxations that were of equal magnitude in rings obtained from both control animals, and those that were hyperchoelsterolemic for 10 weeks (data not shown).

Effect of 17β-estradiol on endothelial NOS via receptor system

Preincubation with 17β-estradiol (10^{-12}-10^{-8} M) for more than 8 hours significantly enhanced the activity of eNOS in both HUVEC and BAEC. A high dose ($10^{-7}$to 10^{-6} M) of 17β-estradiol tended to reduce eNOS activity to the

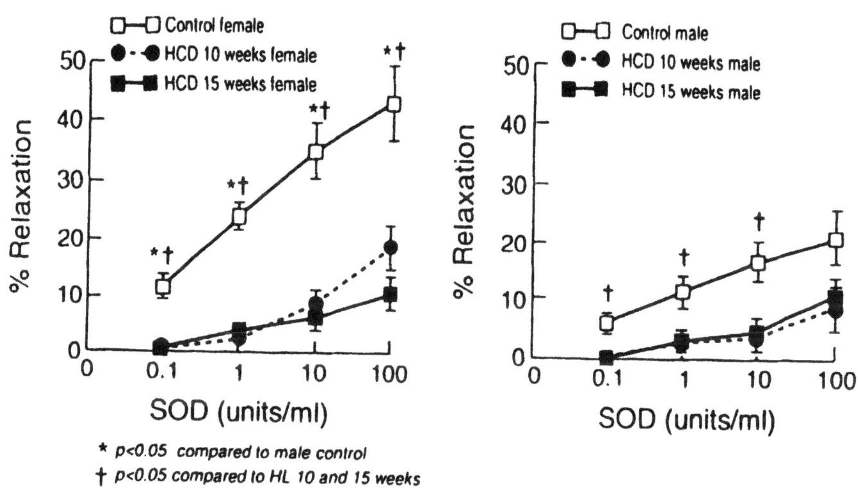

Fig. 3. Basal nitric oxide release. The effects of superoxide dismutase (SOD) on endothelium-intact aortic rings after induction of moderated vascular tone with phenylephrine. The magnitude of relaxation of aortic rings from control female rabbits (left) was significantly greater as compared with the rings from control male rabbits (right). In aortic rings from animals fed a high cholesterol diet (HCD), the magnitude of relaxation of these aortic rings to SOD wa significantly diminished as compared with that in rings from control animals.

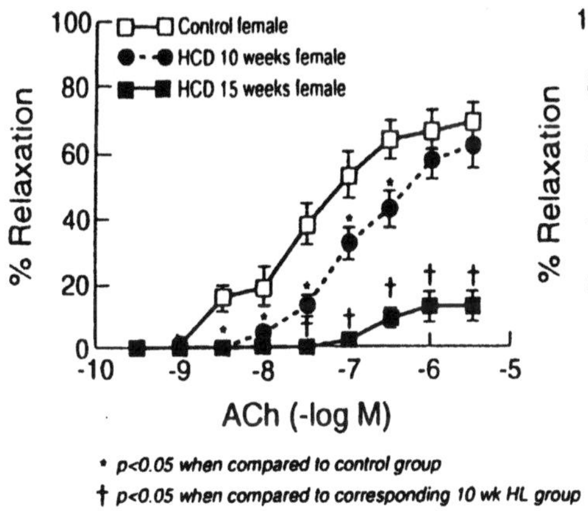

Fig.4. Endothelium-dependent relaxation. The effects of various concentrations of acetylcholine (ACh) on phenylephrine-precontracted endothelium-intact aortic rings obtained from control animals and animals fed a high cholesterol diet (HCD) for 10 or 15 weeks. *; In HCD animals, the magnitude of relaxation of these aortic rings to ACh was diminished as compared with that in rings from control animals ($P<0.05$). †; The ACh response was furhter diminished in HCD 15-weeks animals as compared with HCD 10-week animals ($P<0.05$).

control level (Fig. 5). The release of NO measured with the NO selective electrode showed identical results. NO_2^-/NO_3^- in the culture medium also showed an increase following incubation with physiological concentration of 17β-estradiol. Western blot analysis of the HUVEC and BAEC showed that the anti-monoclonal antibody to eNOS exhibited an effect similar to that of estrogen (data not shown). The estrogen receptor antagonists, tamoxifen and ICI182780, inhibited the effect of 17β-estradiol on eNOS of the endothelial cells by 80%. The inhibitory effect of 17β-estradiol on the activity of eNOS decreased gradually when the culture passage exceeded 10. No significant effect was observed in cells that exceeded 16 passages. Immunocytochemistry showed the existence of an estrogen receptor in both HUVEC and BAEC at less than 5th passage and a marked weakness of that staining in BAEC at more than the 16 th passage (data not shown).

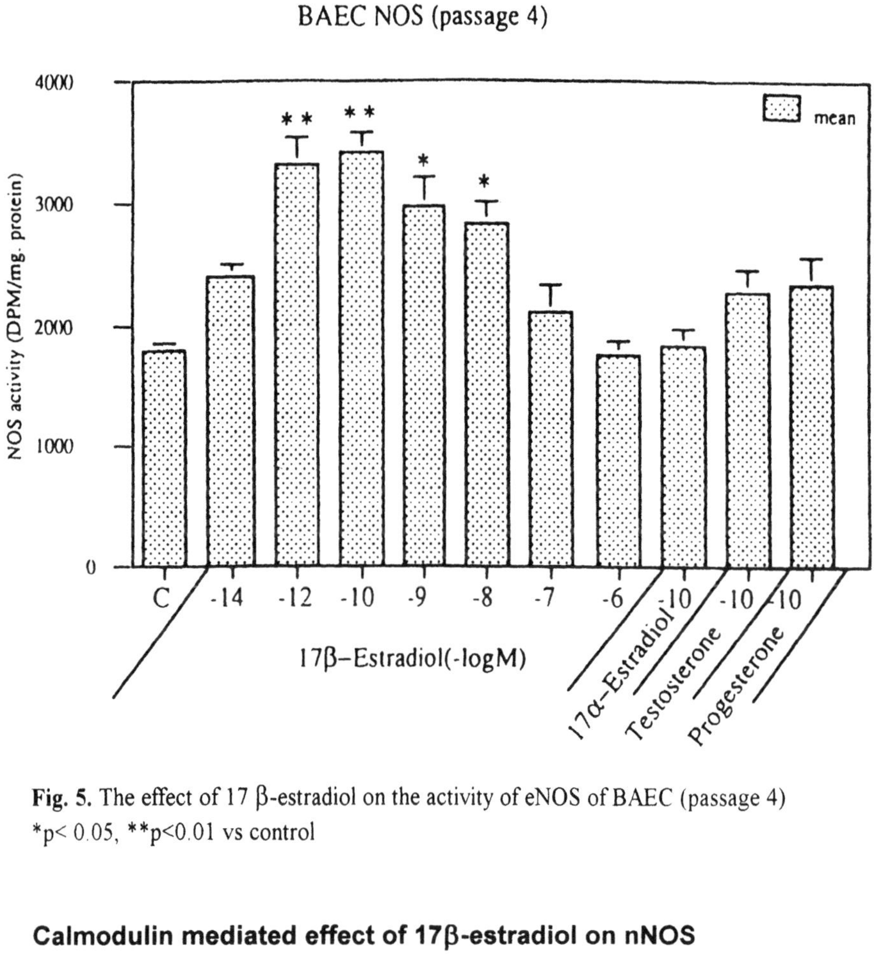

Fig. 5. The effect of 17 β-estradiol on the activity of eNOS of BAEC (passage 4)
*p< 0.05, **p<0.01 vs control

Calmodulin mediated effect of 17β-estradiol on nNOS

A low concentration of 17β-estradiol (10^{-10} to 10^{-8} M) enhanced the activity of crude nNOS in the cytosolic fraction of rabbit cerebellums, while 10^{-9} M 17β-estradiol increased the activity about 20 %. High dose of 17β-estradiol (10^{-5} to 10^{-4} M) augmented the activity about 50 %. The calmodulin antagonists, W7 (10^{-4} M) or calmidazolium ($3x10^{-7}$ M), inhibited nNOS activity by about 30 % in control conditions, and by about 50 % under low dose estrogen conditions and less than 20 % under high dose estrogen conditions. Thus, nNOS activity after W7 or calmidazolium applications was almost identical to that determined in response to any concentration of estrogen. Partially purified nNOS, obtained from the cytosolic fraction by DEAE cellurose columnchromatography, showed a similar enhancement of activity by estrogen, and W7 or calmidazolium inhibited that activity by 90 % in control tissue and that treated with estrogen. Estrogen enhanced the fluorescence of dansyl-calmodulin at low doses, and further augmented that fluorescence at high doses. The estrogen antagonist, tamoxifen,

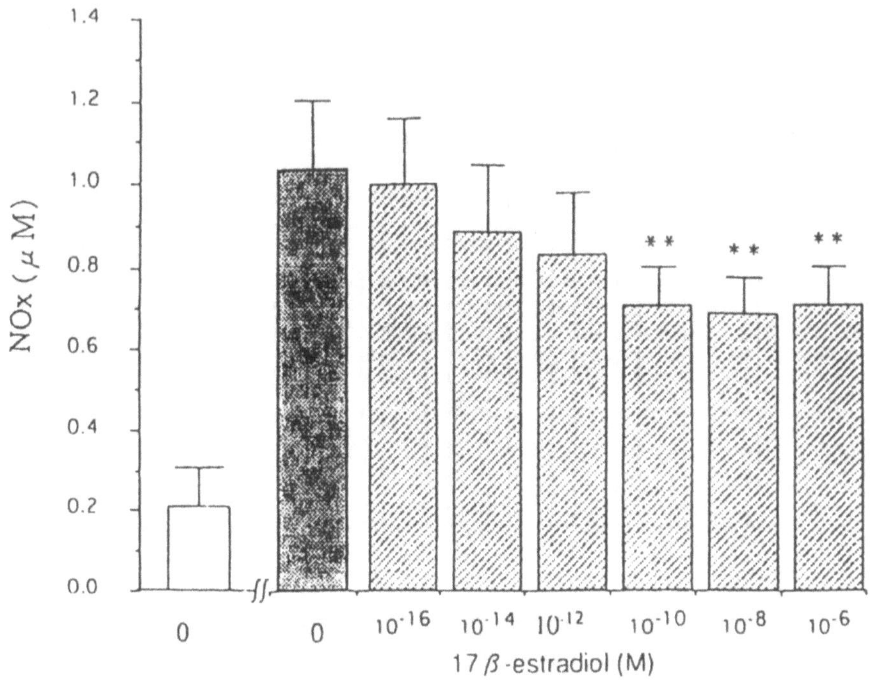

Fig. 6. The effect of 17β-estradiol on the NO_2^-/NO_3^- release from iNOS in J774 cells
*p< 0.05, **p<0.01 vs control

did not affect the activity of nNOS in cerebellum, although it slightly decreased the activity of partially purified nNOS.

The effect of estradiol on inducible nitric oxide synthase

NOx released from J774 cells was measured in cultured medium. Coincubation of 17β-estradiol (>10^{-10} M) decreased the amounts of NO_2^-/NO_3^- in the cultured medium dose-dependently up to 10^{-7} M (Fig.6). Tamoxifen or ICI187780 did not affect the production of NO_2^-/NO_3^- by J774.2 cells, and abolished the effect of 17β-estradiol on the production of NO_2^-/NO_3^- by J774.2 cells when coincubated with 17β-estradiol (data not shown). The effect of 17 β-estradiol was not observed even after 8 hrs after coincubation with IFN-γ and lipoplysaccaride (LPS). In contrast, 1 mM L-NMMA, a competitive inhibitor of iNOS, caused a similar degree of marked iNOS inhibition whether added at 0 time or 8 h after IFN- γ plus LPS.

Discussion

No gender differences were noted in the responsiveness of the aortic rings obtained from the control and hypercholesterolemic animals to ACh, A-23187, or NTG. We also assessed the basal release of NO indirectly by initially inducing moderate active tone and then observing the effects of SOD and L-NMA on changes in that basal tone. There was increased basal release of NO from aortic rings of female rabbits as compared with those from male or oophorectomized rabbits. Despite minimal atherosclerosis in the female that was hyperlipidemic for 10 wks, the basal and stimulated release of NO were impaired to a similar degree in both genders. No significant gender differences, in either the extent of atherosclerosis or in basal and stimulated release of NO, were noted in the hyperlipidemic animals at 15 weeks. These data indicated that hyperlipidemia, rather than the extent of atherosclerosis formation, may be more important altering NO metabolism in the early stages of atherosclerosis formation. The increased basal NO release observed in female aorta may be important in preventing the early stage of atherosclerosis formation.

We also demonstrated that the activity of eNOS was enhanced by physiological concentrations of estrogen. Incubation of HUVEC and BAEC (less than 5th passage) with 17β-estradiol for more than 8 hours enhanced the activity of eNOS. Thus estrogen may affect vascular tone by controlling the activity of eNOS in the endothelium. Preincubation with 17β-estradiol also enhanced the release of NO from the endothelial cells as measured by an NO selective meter and NO_2^-/NO_3^-, metabolites of NO.

Both estrogen receptor antagonists, tamoxifen and ICI182780, inhibited the effect of 17β-estradiol on eNOS by more than 80%. Because tamoxifen is a partial estrogen agonist, we also used the new specific estrogen receptor antagonist, ICI182780. On the other hand, 17β-estradiol may work on eNOS by an other mechanism, such as by enhancing the Ca^{2+}/calmodulin system as we previously reported [9].

Western blot analysis of HUVEC and BAEC using a monoclonal antibody to eNOS, showed an similar effect of estrogen. This probably would occur during protein synthesis or gene transfer. Altough human and bovine eNOS have been identified at the gene level, the existence of an estrogen receptor responsive element remains controversial in endothelial cells of both species. The left-half palindromic sites of an estrogen responsive element (GGTCA) and the right-half sites (TGACC) have been identified in both kinds of cells [12]. Results of previous studies that evaluated the effect of estrogen on eNOS are controversial [13,14]. One study showed that estrogen increased the eNOS activity of BAEC and HUVEC primary cultures, while the other showed that estrogen did not affect the activity of eNOS in BAEC. This discrepancy may have arisen from differences in culture passages or numbers of cell division. In our study, the effect of 17β-estradiol on the activity of eNOS decreased gradually beyond culture passage10, and no significant effect was observed in the cells after 16 passages.

Immunocytochemistry showed the existence of estrogen receptor in cultured HUVEC and BAEC at less than the 5th passage and it became sparse in BAECs at more than the16th passage. The mechanism of the change in the estrogen receptor during repeated passages may depend on the estrogen concentration of the plasma, the medium of the cultured cells, or on the number of cell division times. The number or activity of estrogen receptors in endothelial cells can thus be affected by factors other than estrogen sulfotransferase. The mechanism underlying the change in estrogen receptor warrants study.

Recently, estrogen was reported to decrease concentration of O_2^- from BAEC, and it is a possible mechanism with which estrogen enhances NO action [15]. However, the passages which they used were between 5 th and 15th, so other possibility remains obscure. Numerous mechanisms have been proposed to explain the antiatherosclerotic effect of estrogen [1,16]. However, our finding that estrogen increases eNOS activity via a receptor-mediated system could be a mechanism to explain the effect of estrogen. Its diminution with successive passages suggests the importance of the aging process in the effects of estrogen on vascular cells [17]. The NO-mediated endothelium-dependent relaxation is decreased with age [18]. The existence of the estrogen receptor is likely to be important for the effect of the NO released from endothelium.

Acknowledgment. We wish to thank Norie Kametsuta for her excellent technical assistance. We also appreciate Louis J. Ignarro and Gautam Chaudhuri of the Department of Pharmacology and Ob/Gyn, University of California, Los Angeles, USA for their technical suggestions that assisted in determining the activity of nitric oxide synthase.

References

1. Kannel WB, Hjortland MC, McNamara PM, Gordon T (1976) Menopause and risk of cardiovascular disease: the Framingham Study. Ann Intern Med 85: 447-452
2. Stampfer MJ, Colditz BA (1981) Estrogen replacement therapy and coronary heart disease: a quantitative assessment of the epidemiologic evidence. Prev Med 20: 47-63
3. Gerrity RG (1981) The role of monocyte in atherogenesis: I. Transition of blood-borne monocytes into foam cells in fatty lesions. Am J Pathol 103: 182-190
4. Witzum JL, Simmons D, Steinberg D, Beltz WF (1989) Intensive combination drug therapy of familial hypercholesterolemia with lovastation, probucol and colestipol hydrochloride. Circulation 79: 16-28

5. Lipid Research Clinics Program (1974) Manual of laboratory operations, vol 1, lipid and lipoprotein analysis. U.S. DHEW, publ. no. (NIH) 75-628. Washignton, D.C., U.S. Goverment Printing Office
6. Furchgott RF, Zawadzki JV (1980) The obligatory role of endothelial cells in the relaxation of arterial smooth muscle by acethylcholine. Nature 288: 373-376
7. Hayashi T, Fukuto JM, Ignarro LJ, Chaudhuri G (1992) Basal release of nitric oxide from aortic rings is greater in female rabbits than in male rabbits: implications for atherosclerosis. Proc Natl Acad Sci USA 89: 11259-11263
8. Griscavage JM., Rogers NE, Sherman MP, Ignarro LJ (1993) Inducible nitric oxide synthase from a rat alveolar macrophage cell line is inhibited by nitric oxide. J Immunol 151: 6329-6337
9. Hayashi T, Ishikawa T, Yamada K, Kuzuya M, Naito M, Hidaka H, Iguchi A (1994) Biphasic effect of estrogen on neuronal constitutive nitric oxide synthase via Ca^{2+} -calmodulin dependent mechanism. Biochem Biophys Res Commun 203: 1013-1019
10. Ichimori K, Ishida M, Fukahori H, Nakazawa H, Murakami E (1994) Practical nitric oxide measurement employing a nitric oxide-selective electrode. Rev Sci Instrum 65: 1-5
11. Kanno K, Hirata Y, Emori T, Ohta K, Imai T, Marumo F (1995) L-arginine infusion induces hypotension and diuresis/natriuresis with concomitant increased urinary excretion of nitrate/nitrict and cGMP in man. Clin ExpPharmacol Physiol 19: 619-625
12. Hishikawa K, Nakaki T, Marumo T, Suzuki H, Kato R, Saruta T (1995) Up-regulation of nitric oxide synthase by estradiol in human aortic endothelial cells. FEBS letters 360: 291-293
13. Hassan SS, OharaY, Navas JP, Peterson TE, Dockery S, Harrison DG (1993) Circulation 88: I-80
14. Schray-Utz B, Zeiher AM, Busse R (1993) Circulation 88: I-80
15. Arnal JF, Clamens S, Pechet C, Negre-Salvayres A, Allera C, Girolan J-P, Salvayre R, Bayard F (1996) Ethinylestradiol does not enhance the expression of nitri coxide synthase in bovine endothelial cells but increases the release of bioactive nitric oxide by inhibiting superoxide anion production. Proc Natl Acad Sci USA 93: 4108-4113
16. Collins P, Shay J, Jiang C, Moss J (1994) Nitric oxide accounts for dose-dependent estrogen-mediated coronary relaxation after acute estrogen withdrawal. Circulation 90: 1964-1968
17. Hayashi T, Yamada K, Esaki T, Iguchi A (1995) Estrogen increases endothelial nitric oxide by a receptor mediated system. Biochem Biophys Res Commun 214: 847-855
18. Flavahan NA (1992) Atherosclerosis or lipoprotein-induced endothelial dysfunction. Potential mechanisms underlying reduction in EDRF/nitric oxide activity. Circulation 85: 1927-1938

Part 3

NO Production

Measurement of Exhaled Nitric Oxide of Lung Origin

Ichizo Tsujino, Hideki Shinano, Kenji Miyamoto, Masaharu Nishimura, Yoshikazu Kawakami[1]

SUMMARY. The concentration of nitric oxide ([NO]) in human exhaled air has been repeatedly reported as a promising parameter which reflects inflammatory processes of airway diseases such as bronchial asthma. However, since the exact source of exhaled NO has not been fully understood, we first examined the effects of endotracheal intubation and of breath holding on exhaled [NO] in normal humans. Using a specially-designed gas sampling system, we divided exhaled air into two fractions, the initially exhaled 200ml (Fraction 1; F1) and the remainder (Fraction 2; F2). Before intubation, exhaled [NO] was 15.5±3.6 ppb in F1 and 10.3±2.4 ppb in F2. Compared to these values, [NO] after intubation significantly decreased by 59.2±10.6 (SD)% in F1 and by 54.4±8.0% in F2. Breath holding under endotracheal intubation significantly increased exhaled [NO] only in Fraction 1. In the next study, we developed a nasal continuous negative pressure technique which, we verified, caused closure of the vellum and removed nasal NO contamination, and measured orally exhaled [NO] of healthy humans and of patients with bronchial asthma. In patients with bronchial asthma, exhaled [NO] obtained with nasal continuous negative pressure (CNP) was 20.8±3.4 ppb in Fraction 1, and 19.4±4.2 ppb in Fraction 2, both of which wrewere significantly higher than that of healthy humans. These results indicate those about 55-60% of exhaled NO is derived from the upper airways in healthy humans when they breathe wearing a noseclip, and that production of NO in intrathoracic airways is augmented in patients with asthma compared to healthy humans.

KEY WORDS: Human, endotracheal intubation, bronchial asthma, Intrathoracic airways

---

1 First Department of Medicine, Hokkaido University School of Medicine. Sapporo, Hokkaido, 060-8638 Japan

Introduction

Nitric oxide has been repeatedly detected in exhaled air of humans [1, 2, 3] since the first report by Gustafsson and his colleagues in 1991 [4]. The concentration of NO ([NO]) in exhaled air has been reported to be high in patients with bronchial asthma [5, 6, 7] and bronchiectasis [8], and to be low in smokers [9, 10] and hypertensive subjects [11]. However, in contrast to some investigations arguing that the major source of exhaled NO is the upper respiratory tract, in particular, the nasal cavity [12, 13], others believe that exhaled NO sampled when subjects are wearing a noseclip mostly comes from lower airways or from alveoli [1, 2]. Considering the uncertainty on this issue, it would be of importance to examine the effects of endotracheal intubation and of the respiratory pattern on exhaled [NO] to gain insight into the source of exhaled NO. Furthermore, if there was a significant contamination of orally exhaled gas by nasal NO, elimination of such contamination would be vital to learn what is happening exclusively in the lungs. In the present study, we developed a new method to remove nasal NO contamination. and examined its validity. Subsequently, we measured exhaled [NO] of patients with asthma using the technique and made comparison with that of healthy humans.

Methods

Study 1: Source of exhaled NO in healthy humans

Subjects:
Six healthy male volunteers aged 19-35 years served as subjects for this study with written informed consent. They included 1 current smoker whose smoking index was 2 pack year, and was requested not to smoke at least for six hours prior to the study. None of the subjects had had any complaints of respiratory symptoms for the previous three months. This study was approved by the ethics committee of the Hokkaido University School of Medicine.

Experimental setup (Fig 1):
The subject sat on a chair and, while wearing a noseclip, breathed through a mouthpiece which was connected to a T-valve. Inhaled gas was prepared by blending oxygen, nitrogen and also carbon dioxide, and the concentration of NO in the inhaled gas was confirmed to be less than 1 part per billion (ppb) before each experiment. Exhaled gas was collected through a T-valve into the gas sampling system in the expiratory line which consisted of a hotwire flow meter (RF-2, Minato Ikagaku Corporation, Osaka) and the three sequential outlets connected by NO-impermeable bags made from polyvinyl fluorine film (Tedlar bag, GL Sciences. Tokyo). Each outlet had an electromagnetic valve which shut or opened according to the signal from the system controller. The first bag was designed to collect about

140ml of initially exhaled gas, which was equivalent to the dead space volume of the system. The second bag was designed to collect about 200ml of subsequently exhaled gas and the third was for the rest of the exhaled gas. The gas sampled in the second bag was labeled as Fraction 1 and that in the third one as Fraction 2. We assumed in this experiment that Fraction 1 would include more gas derived from airways and less from alveoli than Fraction 2. The [NO] in collected gas was measured by a chemiluminescence NO analyzer (Model 42S, Thermo Environmental Instruments Inc., Franklin, MA), and two-point calibration was done at zero and 75 ppb before each experiment using the calibration gas generators (Model 102 and Model 97S, Thermo Environmental Instruments Inc., Franklin. MA).

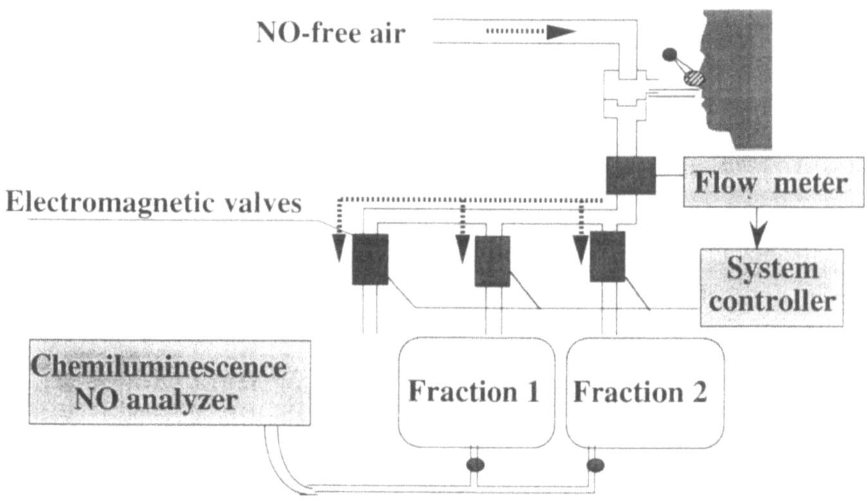

Fig. 1. Experimental setup

Experimental protocol:
To evaluate the influence of NO from the upper airways on exhaled [NO], we compared exhaled [NO] before and after endotracheal intubation in 6 subjects. Subjects were first instructed to take normal tidal volume breaths at a fixed rate of 15 breaths/min, and exhaled [NO] was measured while they breathed wearing a noseclip. Subsequently, we locally anesthetized the subject's pharyngolarynx by spraying 2% xylocaine (5-10ml) and inserted the 8.0 Fr intubation tube with inflating cuff (Laboratoire Portex S. A., Berck, France), which should completely abolish gas mixture between extrathoracic upper airways and the lungs. After this

procedure, we measured exhaled [NO] while subjects were taking normal breaths (n=6) and also after 0, 2 and 8 second breath holding (n=4).

Study 2: Measurement of exhaled NO from intrathoracic airways in patients with bronchial asthma

Subjects:
Ten patients who had been diagnosed as bronchial asthma were recruited. Their ages varied from 17 to 63 years, and none of the subjects had any history of other respiratory or cardiovascular diseases. Measurement of exhaled [NO] was carried out when their symptoms were not completely controlled so that there were variable levels of expiratory wheezing on auscultation in all the subjects on the experimental day. Six of the patients were receiving inhaled or oral glucocorticoids. As control subjects, six healthy male volunteers aged 19-25 years served with written informed consent. None of them had had any complaints of respiratory symptoms for the previous three months. This study was approved by the Ethics Committee of the Hokkaido University School of Medicine.

Removal of nasal NO contamination:
In this study, we used the same gas sampling system which was described in study 1. We first developed a nasal continuous negative pressure (nasal CNP) technique to close the vellum and to remove nasal NO contamination, in which we applied negative pressure using an airtight nasal mask connected by a suction system equipped on the wall of each room in clinical wards. Pressure in the suction line was maintained at -5 cmH_2O throughout the measurement. To confirm that nasal CNP causes closure of the vellum, we took lateral X-ray films of the upper airways in one subject, and assured its closure by a lifted-up soft palate both in the inspiratory and expiratory phases. Additionally, we also measured tidal volume (VT), vital capacity (VC) and CO_2 concentration in Fraction 1 and 2 while the subjects were breathing first with a noseclip and then with nasal CNP. Since there was no significant difference in these parameters between the two methods, nasal CNP was again proved to close the vellum.

Exhaled [NO] in healthy subjects and in patients with bronchial asthma:
Exhaled [NO] was measured in 6 healthy volunteers and in 10 patients with bronchial asthma first while they were breathing with a noseclip, and then with nasal CNP. Exhaled [NO] obtained with the two methods was compared between the two groups. To estimate what percentage of NO is derived from the upper airways, the reduction rate of [NO] caused by nasal CNP was calculated for both fractions. The reduction rate was defined as follows;

Reduction rate (%) = {([NO]noseclip - [NO]nasal CNP) / [NO]noseclip} x 100, where [NO]noseclip refers to [NO] obtained with a noseclip and [NO]nasal CNP to [NO] with nasal CNP

Statistics

All data were expressed as means $\pm$ SE unless otherwise specified. For the statistical analysis of difference in [NO], Student's paired t-test was applied, and a p-value of less than 0.05 was considered significant.

Results

study 1.

Before intubation, [NO] in Fraction 1 and 2 was 15.5 ± 3.6 ppb and 10.3 ± 2.4 ppb, respectively, and the value in Fraction 1 was significantly higher than that in Fraction 2. After intubation, [NO] in both fractions significantly decreased to 5.7 ± 1.2 in Fraction 1 and 4.8 ± 1.0 ppb in Fraction 2 (Fig. 2). The reduction of exhaled [NO] was $59.2 \pm 10.6\%$ and $54.4 \pm 8.0\%$ in Fraction 1 and 2, respectively. Breath holding under endotracheal intubation significantly increased exhaled [NO] only in Fraction 1, from 6.4 ± 1.2 ppb without breath holding to 13.8 ± 3.1 ppb with 8 sec breath holding ($p<0.05$) as is shown in Fig. 3.

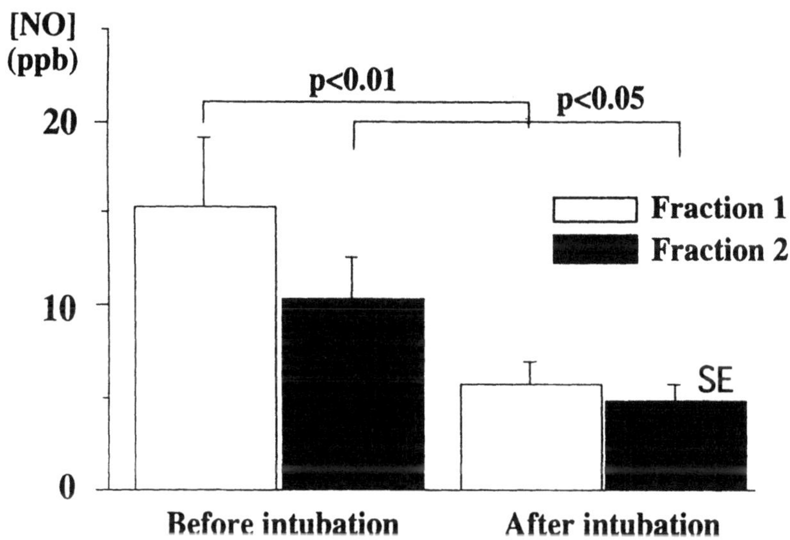

Fig. 2. Exhaled [NO] before and after endotracheal intubation
Before intubation, [NO] in Fraction 1 was 15.5 ± 3.6 ppb and was significantly higher than 10.3 ± 2.4 ppb in Fraction 2. After intubation, [NO] in both fractions significantly decreased to 5.7 ± 1.2 ppb in Fraction 1 ($-59.2 \pm 10.6\%$) and to 4.8 ± 1.0 ppb in Fraction 2 ($-54.4 \pm 8.0\%$).
Fraction 1; initially exhaled 200ml of exhaled air, Fraction 2; the rest of exhaled air.

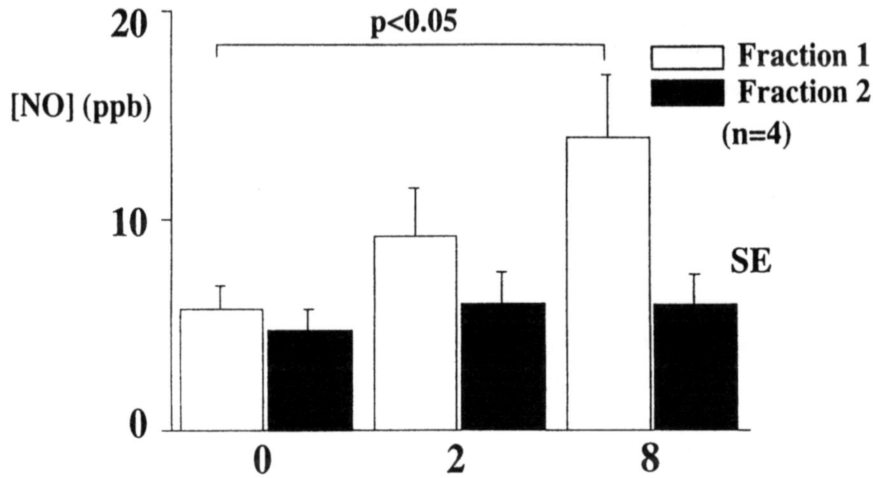

Fig. 3. Effect of breath holding under endotracheal intubation
[NO] in Fraction 1 increased with 2 and 8 sec breath holding compared to that with 0 sec breath holding.
Fraction 1; initially exhaled 200ml of exhaled air, Fraction 2; the rest of exhaled air.

study 2.

In 10 patients with bronchial asthma, exhaled [NO] was 37.9 ± 7.0 ppb in Fraction 1 and 32.7 ± 5.9 ppb in Fraction 2. Although the concentration slightly decreased by nasal CNP, the values in both fractions were still significantly higher compared to those of healthy humans (Fig. 4). Percent reduction of exhaled [NO] caused by nasal CNP was calculated, and was $23.6 \pm 14.4\%$ in Fraction 1 and $22.7 \pm 14.4\%$ in Fraction 2 in patients with bronchial asthma. Although the difference did not reach statistical significance, both of the values were lower than those of healthy volunteers which was $44.5 \pm 11.3\%$ in Fraction 1 and $33.7 \pm 10.9\%$ in Fraction 2.

Discussion

In this study, we first demonstrated, using a specially-designed technique for sampling exhaled air, that [NO] was consistently higher in Fraction 1 than in Fraction 2 whenever when the subjects breathed wearing a noseclip or under intubation. Moreover, the incremental effect of the breath holding maneuver on

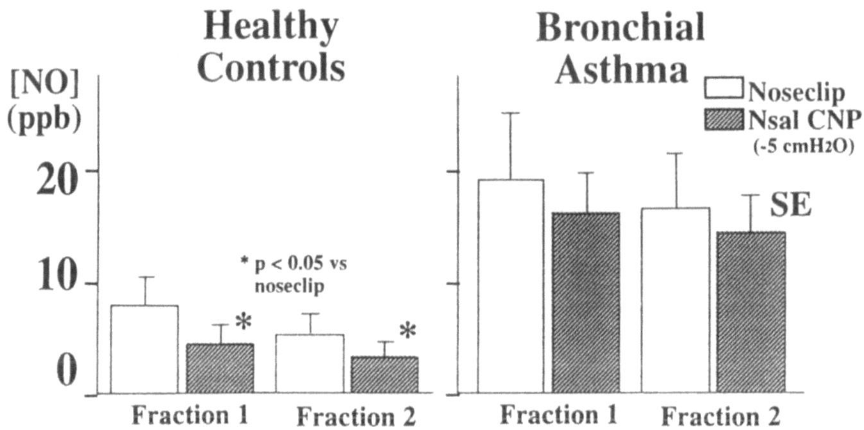

Fig. 4. Exhaled [NO] of healthy humans and of patients with bronchial asthma
Compared with the values of healthy humans, exhaled [NO] in patients with asthma was consistently higher irrespective of the measuring method. Percent reduction of exhaled [NO] caused by nasal CNP was calculated, and the values of asthmatic patients (Fraciton 1: 23.6 ± 1 4.4%, Fraction 2: 22.7 ± 14.4%) were lower than those of healthy volunteers (Fraction 1: 44.5 ± 11.3%, Fraction 2: 33.7 ± 10.9%).
Fraction 1; initially exhaled 200ml of exhaled air, Fraction 2; the rest of exhaled air.

exhaled [NO] was significant only in Fraction 1. These results indicated that [NO] detected in exhaled air of healthy humans was derived mainly from the airway, not from the alveoli. Secondly, we demonstrated that, even if subjects wore a noseclip, a substantial part of exhaled NO was derived from upper airways, including the nasal cavity, in normal humans, since exhaled [NO] sampled through the intubated bronchial tube was lower by 59.2 ± 10.6% in Fraction 1 and 54.4 ± 8.0% in Fraction 2 than that sampled without intubation.

It must be noted that wearing a noseclip does not eliminate the effect of upper airways, including the nasal cavity, on exhaled [NO]. In most previous studies, whatever method was used for the measurement of exhaled [NO], the assumption was made that [NO] in exhaled air represented something within the lungs as long as subjects wore a noseclip. Although this study demonstrated that less than 50% of NO in exhaled air was certainly derived from the lungs, care must be taken when interpreting the data of exhaled [NO]. This issue was also fully discussed in our previous report [14].

More recently, a new technique was proposed by Silkoff [15] and Kharitonov [16], in which they orally applied positive expiratory pressure using a resistance load in an expiratory line, and successful removal of nasal NO was shown in the reports. This technique is definitely superior to the previous methods because of its simple and noninvasive nature. However, since this method can accompany influence of positive pressure on the pulmonary circulation or mechanics, or inhalation of nasal NO during the inspiratory phase, we believe nasal CNP technique could be an another choice to remove nasal NO contamination without such influence.

Some recent studies have provided evidence that production of NO is increased in lower airways in patients with bronchial asthma. In the present study, we confirmed that exhaled [NO] in patients with bronchial asthma was higher than in healthy humans even after contamination of nasal NO was eliminated by the nasal CNP method. Moreover, this study demonstrated that reduction rate of [NO] caused by the nasal negative pressure method was lower in patients with bronchial asthma than in healthy humans in both fractions. Since the reduction rate should reflect the extent of nasal NO contribution in exhaled [NO], the lower rate suggests relatively low contribution of nasal NO in patients with bronchial asthma.

In conclusion, NO detected in exhaled air of healthy humans comes mainly from the airway, not from the alveoli. However, it must be noted that, when subjects breathe wearing a noseclip, significant contamination by NO from the upper airways occurs in orally exhaled NO. Nasal CNP through an airtight nasal mask is a simple and noninvasive way to remove such contamination. Finally, we confirmed that intrathoracic production of NO is increased in patients with bronchial asthma compared with healthy humans.

References

1. Borland C, Cox Y, Higenbottam T (1993) Measurement of exhaled nitric oxide in man. Thorax 48: 1160-1162
2. Persson MG, Wiklund NP, Gustafsson LE (1993) Endogenous nitric oxide in single exhalations and the change during exercise. Am Rev Respir Dis 148: 1210-1214
3. Iwamoto J, Pendergast DR, Suzuki H, Krasney JA (1994) Effect of graded exercise on nitric oxide in expired air in humans. Resp Physiol 97: 333-345
4. Gustafsson LE, Leone AM, Persson MG, Wiklund NP, Moncada S (1991) Endogenous nitric oxide is present in the exhaled air of rabbits, guinea pigs and humans. Biochem Biophys Res Commun 181: 852-857
5. Alving K, Weitzberg E, Lundberg JM (1993) Increased amount of nitric oxide in exhaled air of asthmatics. Eur Respir J 6: 1368-1370
6. Kharitonov SA, Yates D, Robbins RA, Logan-Sinclair R, Shinebourne EA, Barnes PJ (1994) Increased nitric oxide in exhaled air of asthmatic patients. Lancet 343: 133-135

7 Massaro AF, Gaston B, Kita D, Fanta C, Stamler JS, Drazen MJ (1995) Expired nitric oxide levels during treatment of acute asthma. Am J Respir Crit Care Med 152: 800-803
8 Kharitonov SA, Wells AU, O'Connor BJ, Cole PJ, Hansell DM, Logan-Sinclair RB, Barnes PJ (1995) Elevated levels of exhaled nitric oxide in bronchiectasis. Am J Respir Crit Care Med 151: 1889-1893
9 Persson MG, Zetterstrom O, Agrenius V, Ihre E, Gustafsson LE (1994) Single-breath nitric oxide measurements in asthmatic patients and smokers. Lancet 343: 146-147
10. Kharitonov SA, Robbins RA, Yates D, Keatings V, Barnes PJ (1995) Acute and chronic effects of cigarette smoking on exhaled nitric oxide. Am J Respir Crit Care Med 152: 609-612
11. Shilling J, Holzer P, Guggenbach M, Gyurech K, Marathia K, Geroulanos S (1994) Reduced endogenous nitric oxide in the exhaled air of smokers and hypertensives. Eur Respir J 7: 467-471
12. Lundberg JON, Weitzberg E, Nordvall SL, Kuylenstierna R, Lundberg JM, Alving K (1994) Primarily nasal origin of exhaled nitric oxide and absence in Kartagener's syndrome. Eur Respir J 7: 1501-1504
13. Gerlach H, Rossaint R, Pappert D, Knorr M, Falke KJ (1994) Autoinhalation of nitric oxide after endogenous synthesis in nasopharynx. Lancet 343 :518-519
14. Tsujino I, Miyamoto K, Nishimura M, Shinano H, Makita H, Saito S, Nakano T, Kawakami Y (1996) Production of nitric oxide (NO) in intrathoracic airways of normal humans. Am J Respir Crit Care Med 154: 1370-1374
15. Silkoff PE, McClean P, Slutsky A, Furlott H, Hoffstein E, Wakita S, Chapman K, Szalai J, Zamel N (1997) Marked flow-dependence of exhaled nitric oxide using a new technique to exclude nasal nitric oxide. Am J Respir Crit Care Med 155: 260-267
16. Kharitonov SA, Barnes PJ (1997) Nasal contribution to exhaled nitric oxide during exhalation against resistance or during breath holding. Thorax 52: 540-544

Induction of Tetrahydrobiopterin Synthesis in Cardiac Myocytes

Yoshiyuki Hattori, Kikuo Kasai[1]

SUMMARY. Cardiac myocytes are now known to express the high-capacity inducible isoform of nitric oxide (NO) synthase (iNOS). Induction of iNOS by soluble inflammatory mediators, including cytokines, causes a marked depression in myocyte contractile responsiveness to b-adrenergic agonists and has been implicated as a contributor to the pathogenesis of heart failure. Since tetrahydrobiopterin (BH4) is an essential cofactor for NO formation, we investigated whether BH4 synthesis is required for cytokine-induced NO production in cultured rat cardiac myocytes. Activation of NO formation by cytokines in cardiac myocytes requires transcriptional induction of the genes that encode iNOS and GTP cyclohydrolase I (GTPCH), the first and rate-limiting enzyme in *de novo* BH4 synthesis. Given that nuclear factor κB (NF-κB) mediates the induction of iNOS gene expression in various cell types, the role of NF-kB in the induction of iNOS in cytokine-stimulated rat neonatal cardiac myocytes was assessed by examining the effects of pyrrolidine dithiocarbamate (PDTC), an inhibitor of NF-κB activation, on iNOS mRNA expression and subsequent NO production. The effects of PDTC on GTPCH mRNA expression and pterin synthesis were also examined. We here demonstrate that BH4 synthesis is an absolute requirement for induction of NO synthesis by cytokines in cardiac myocytes. We show that PDTC inhibited in a dose-dependent manner both NO and BH4 synthesis induced by a combination of interleukin-1α (IL-1) and interferon-γ (IFN) and that PDTC also prevented the accumulation of iNOS and GTPCH mRNAs induced by IL-1 and IFN. The induction of both genes necessary for NO synthesis in cardiac myocytes appears to be regulated, at least in part, by a common mechanism: NF-κB activation. Our findings also suggest that regulation of pterin synthesis may be an important target for pharmacologic interventions for NO overproduction within the myocardium in cytokine-related cardiac dysfunction.

---

[1] From the Department of Endocrinology, Dokkyo University School of Medicine, Mibu, Tochigi 321-0293, Japan

KEY WORDS: nitric oxide (NO), nitric oxide synthase, cardiac myocyte, tetrahydrobiopterin (BH4), GTP cyclohydrolase I (GTPCH)

Introduction

The role of immune mechanisms in cardiac disease is receiving increasing recognition [1,2]. Proinflammatory cytokines are released locally in various conditions associated with myocardial inflammation, including cardiac allograft rejection, myocardial infarction, myocarditis, and idiopathic cardiomyopathies [1-3]. Clinical and experimental data strongly support the view that cytokines exert complex effects on myocardial contractile function. One of the mechanisms responsible for cytokine-induced contractile dysfunction seems to be the induction of the high capacity inducible isoform of nitric oxide (NO) synthase (iNOS) in cellular constituents of ventricular muscle including cardiac myocytes [4-6].

Tetrahydrobiopterin (BH4) is an essential cofactor of all isoforms of NO synthase (NOS) [7]. It was first demonstrated that an increase in BH4 levels was crucial for cytokine-induced NO production in murine fibroblasts [8]. The causal relationship between increased BH4 levels and cytokine-induced increases in NO production was extended to human umbilical vein endothelial cells [9], rat vascular smooth muscle cells [10], a murine macrophage cell line [11], rat glial cells [12], and human thyroid cells [13]. These data suggest that cytokines can alter iNOS activity, not only through induction of NOS protein, but also by influencing co-factor availability. Moreover, recent studies have demonstrated that inflammatory cytokines selectively increase NOS activity in human umbilical vein endothelial cells by increasing BH4 levels [14,15]. Interestingly, it has been suggested that under conditions where BH4 production is attenuated, endothelial NOS from canine coronary arteries may provide H_2O_2 which leads to hydroxyl radical production and oxidative tissue damage [16]. Thus, a close and important relationship between NO and BH4 synthesis has been demonstrated in various cell types. However, the extent of this relationship and its impact on myocardial function remain to be investigated.

The induction of iNOS is required for the increase in NO production in response to immunostimulants in cardiac myocytes, and this process is thought to be regulated predominantly at the level of transcription [17]. Regulation of gene transcription by immunostimulants is mediated by various ubiquitous transcription factors in many different cellular systems. Prominent among these factors is nuclear factor κB (NF-κB), which has been shown to mediate the induction of iNOS in several NO-producing cell types on the basis of studies with iNOS gene promoter constructs [18,19] or with inhibitors of NF-κB [20-22] . However, the mechanism, including the role of NF-κB, by which the iNOS induction signal is transduced from the membrane to the nucleus of cardiac myocytes is remains unclear.

The synthesis of BH4 occurs via two distinct pathways: a *de novo* synthetic pathway in which guanosine triphosphate (GTP) serves as a precursor, and a salvage pathway based on pre-existing dihydropterins [23]. GTP cyclohydrolase I (GTPCH), the first and rate-limiting enzyme in the *de novo* synthesis of BH4, catalyzes the formation of dihydroneopterin triphosphate from GTP. Subsequently, dihydroneopterin triphosphate is converted to BH4 by the actions of 6-pyruvoyltetrahydropterin synthase and sepiapterin reductase, with the concomitant formation of tetrahydropterin intermediates. Immunostimulant-evoked BH4 synthesis occurs with a time course that parallels that of coinduced NO synthesis and is preceded by an increase in GTPCH mRNA [24]. The intracellular concentration of BH4 appears to be rate limiting for NO synthesis [25]. Thus, the cytokine-induced increases in BH4 concentration and NO production in cardiac myocytes may be mediated by a common mechanism such as activation of NF-κB.

We have now investigated the role of NF-κB in iNOS mRNA expression and subsequent NO production as well as GTPCH mRNA expression and pterin synthesis by cytokines in rat cardiac myocytes.

Materials and methods

Cell culture

Primary cultures of rat neonatal cardiac myocytes were prepared as previously described [26,27]. Briefly, cardiac ventricles from 2- or 3-day-old Wistar rats were minced and dissociated with 0.1% collagenase. After dispersed cells were incubated for 60 min, nonattached cardiocytes were collected and seeded into 96-well plates for nitrite assay and into 10-cm dishes for biopterin assay and RNA preparation. We routinely obtained enriched cultures containing more than 95% myocytes, as assayed by immunofluorescence staining with a myosin heavy chain antibody.

Nitrite production

Nitrite production, an indicator of NO synthesis, was measured in the supernatants of cardiac myocytes as previously described [25]. In studies which assessed the effect of drugs on induced nitrite production, drugs were added simultaneously with the inducing agents. Cell viability, assessed by measuring cell respiration with MTT [3- (4,5 - dimethylthiazol-2-yl) - 2,5-diphenyltetrazolium bromide] [28] was not significantly diminished, even by the highest concentrations of drugs tested: 2,4-diamino-6-hydroxypyrimidine (3 mM), sepiapterin (300 μM), methotrexate (30 μM), and tetrahydrobiopterin (100 μM). Nitrite was measured by adding 100 μl of Griess reagent (1% sulfanilamide and 0.1% naphthylethylenediamide in 5% phosphoric acid) to 100 μl samples of cell culture medium. The optical density at 550 nm (OD_{550}) was measured with a microplate reader. Nitrite concentrations were calculated by comparison with the OD_{550} of standard solutions of sodium nitrite prepared in culture medium.

Biopterin assay

Total cellular biopterin (biopterin plus BH2 and BH4) was measured after acidic oxidation of the reduced forms of biopterin with iodine as previously described [29]. Briefly, cells were treated with 0.2 M perchloric acid (PCA) and oxidized by exposure to 0.2 M PCA containing 0.2% I_2 and 0.4% KI for 1 h at room temperature in the dark. Ascorbate (2%) was added to remove residual free I_2 and the mixture was centrifuged for 10 min at 10,000 x *g*. The amount of biopterin in the supernatant was quantitated by C_{18} reversed phase high-performance liquid chromatography with fluoresence detection using authentic biopterin as a standard. Protein concentration was measuered by the Lowry method [30] with bovine serum albumin as a standard.

Measurement of mRNA levels for iNOS and GTPCH by RT-PCR

RNA was extracted from the cells by a modified guanidinium isothiocyanate method (RNAzol TM, Cinna/Biotecx, Houston, TX, USA). Separate aliquots of total RNA (1 μg) were reverse-transcribed into cDNA and run through the PCR, as described previously [24]. Primers used were: NOS forward 21-mer, 5'-CTGCAGGTCTTTGACGCTCGG-3'; NOS reverse 21-mer, 5'-GTGGAACACAGGGGTGATGCT-3'; GTPCH forward 21-mer, 5'-GGATACCAGGAGACCATCTCA-3'; GTPCH reverse 21-mer, 5'-TAGCATGGTGCTAGTGACAGT-3'. Products were analyzed on 1.5% agarose gels and photographed under ultraviolet light after being stained with ethidium bromide. PCR amplification was performed using a DNA PCR kit (Perkin Ermer Cetus, Norwalk, CT) according to the following schedule: denaturation, annealing and elongation at 95°, 55° and 72°C for 30 sec, 30 sec and 1 min, respectively, for 30 cycles. To ensure that equal amounts of reverse-transcribed RNA were applied to the PCR reaction, the parallel expression of glyceraldehyde-3-phosphate-dehydrogenase (GAPDH) [31] mRNA was tested after 30 cycles of amplification.

Materials

Recombinant murine interleukin-1α and interferon-γ were obtained from Genzyme (Cambrigde, MA, USA). Sepiapterin was from Alexis (San Diego, CA, USA). Unless stated otherwise, all other compounds were purchased from Sigma (St. Louis, MO, USA).

Results

The total biopterin content (BH4 and more oxidized states) of untreated cardiac myocytes was below our limit of detection. However, treatment with interleukin-1α and interferon-γ (IL-1/IFN) caused a significant rise in biopterin levels and induced NO synthesis (Fig. 1). 2,4-Diamino-6-hydroxypyrimidine (DAHP), a selective inhibitor of GTP cyclohydrolase I (GTPCH, the rate-limiting enzyme for

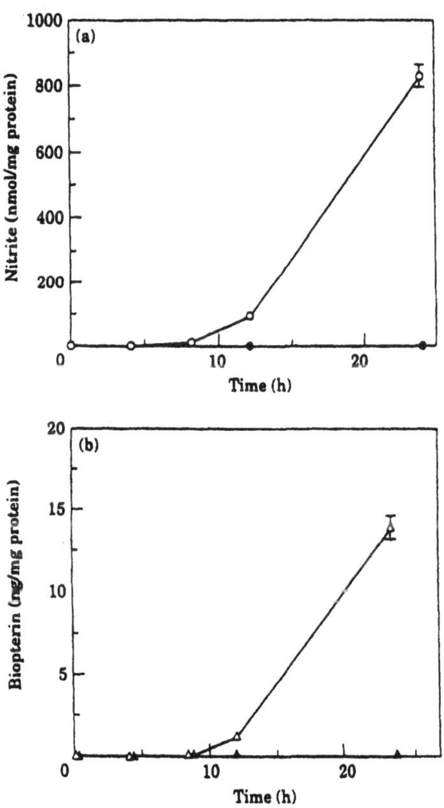

Fig. 1. Time course of nitrite (a) and biopterin (b) production induced by IL-1 (2 ng/ml) and IFN (100 U/ml) in rat cardiac myocytes. Nitrite in the culture medium (a) and cellular biopterin (BH4 and more oxidized species) (b) were assayed for cells incubated for the indicated times in the absence (closed symbols) or presence (open symbols) of IL-1/IFN. Data are means ± SEM of four experiments.

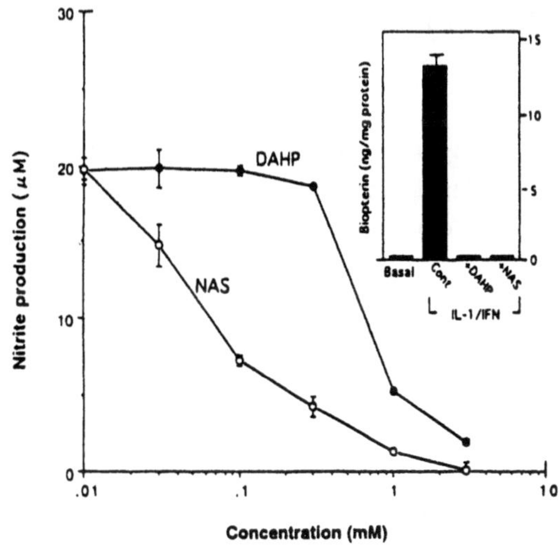

Fig. 2. Modulators of tetrahydrobiopterin (BH4) synthesis on IL-1 and IFN (IL-1/IFN)-activated nitrite synthesis in cardiac myocytes. Effect of DAHP or NAS on nitrite accumulation was assayed after 24-h exposure to IL-1/IFN in combination with different concentrations of 2,4-diamino-6-hydroxypyrimidine (DAHP) and N-acetylserotonin (NAS). Inset, DAHP or NAS prevented increases in biopterin content induced by IL-1/IFN. Each point and bar represents the mean of four values ± s.e.

de novo BH4 synthesis), completely abolished the elevation in biopterin levels induced by IL-1/IFN. DAHP also caused a concentration-dependent inhibition of IL-1/IFN-induced NO synthesis. Similarly, N-acetylserotonin (NAS), an inhibitor of the BH4 synthetic enzyme sepiapterin reductase, blocked increases in biopterin levels as well as NO synthesis induced by IL-1/IFN (Fig.2). Sepiapterin, substrate for BH4 synthesis via the pterin salvage pathway, prevented this inhibition by DAHP or NAS, and this effect was blocked by methotrexate. Sepiapterin, and to a lesser extent BH4, dose-dependently enhanced IL-1/IFN-induced NO synthesis (Fig.3), suggesting that the concentration of BH4 limits the rate of NO production by IL-1/IFN-activated cardiac myocytes.

We next assessed the role of NF-κB in the induction of iNOS in cytokine-stimulated rat neonatal cardiac myocytes by examining the effects of pyrrolidine dithiocarbamate (PDTC), an inhibitor of NF-κB activation, on the abundance of iNOS mRNA and NO synthesis. The effects of PDTC on GTPCH mRNA abundance and biopterin synthesis were also investigated. PDTC inhibited in a dose-dependent manner both NO and BH4 synthesis induced by IL-1/IFN (Fig.4). Whereas iNOS and GTPCH mRNAs were not present in untreated cardiac myocytes, exposure of cells to IL-1/IFN triggered the appearance of both

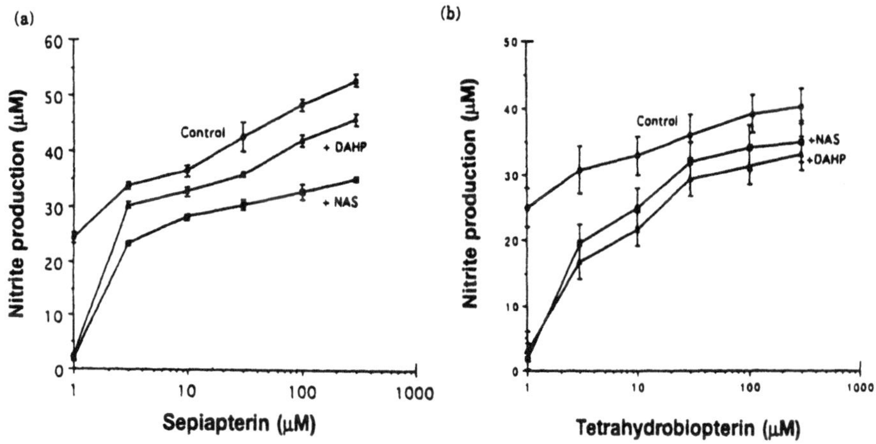

Fig. 3. Influence of sepiapterin and tetrahydrobiopterin concentrations on IL-1 and IFN (IL-1/IFN)-induced nitrite production by rat cardiac myocytes. Effect of sepiapterin (a) or tetrahydrobiopterin (b) on nitrite accumulation after a 24-h exposure to IL-1/IFN alone (Control) or in combination with 2,4-diamino-6-hydroxypyrimidine (DAHP: 3 mM) or N-acetylserotonin (NAS: 1 mM). Each point represents the mean of four values ± s.e.

transcripts. Induction of iNOS and GTPCH mRNAs by IL-1/IFN was abolished by actinomycin D. The expression of iNOS mRNA induced by IL-1/IFN was not influenced by simultaneous treatment with cycloheximide, while the induction of GTPCH mRNA by IL-1/IFN was rather enhanced by cycloheximide. Control PCR experiments demonstrated equivalent expression of the GAPDH gene in all samples (Fig. 5a). We here examined the effects of PDTC on the induction of iNOS and GTPCH mRNAs. PDTC (12.5 to 100 μM) dose-dependently inhibited the IL-1/IFN-induced increase in the abundance of both iNOS and GTPCH mRNAs (Fig.5b). For quantification, we normalized the iNOS and GTPCH signals relative to the corresponding GAPDH signal from the same RNA, and the ratio (iNOS/GAPDH and GTPCH/GAPDH) was shown in lower pannel of Fig.5b. This inhibitory effect of PDTC on the induction of iNOS and GTPCH mRNAs is consistent with the inhibition of NO and BH4 synthesis, respectively.

Discussion

Although immune mechanisms and cytokine-induced NO synthesis in cardiac diseases have been the focus of much recent interest, the cellular mechanisms by which cytokines induce NO synthesis in cardiac myocytes remains obscure. In the

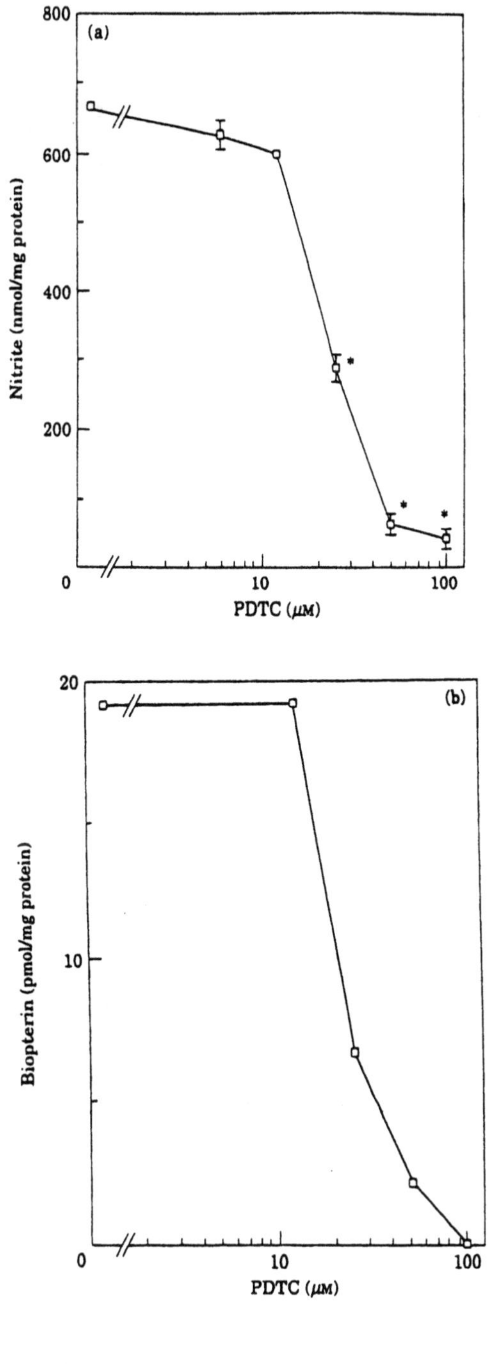

Fig. 4. Effects of PDTC on nitrite and pterin synthesis induced by IL-1/IFN. Rat cardiac myocytes were incubated for 24 h with IL-1/IFN in the absence or presence of the indicated concentrations of PDTC, after which nitrite accumulation in the culture medium and cellular biopterin (BH4 and more oxidized species) contents were determined. Data for nitrite are means ± SEM of four experiments. $^{**}P < 0.01$ compared with control (no PDTC). Data for biopterin are means of two experiments that yielded similar results.

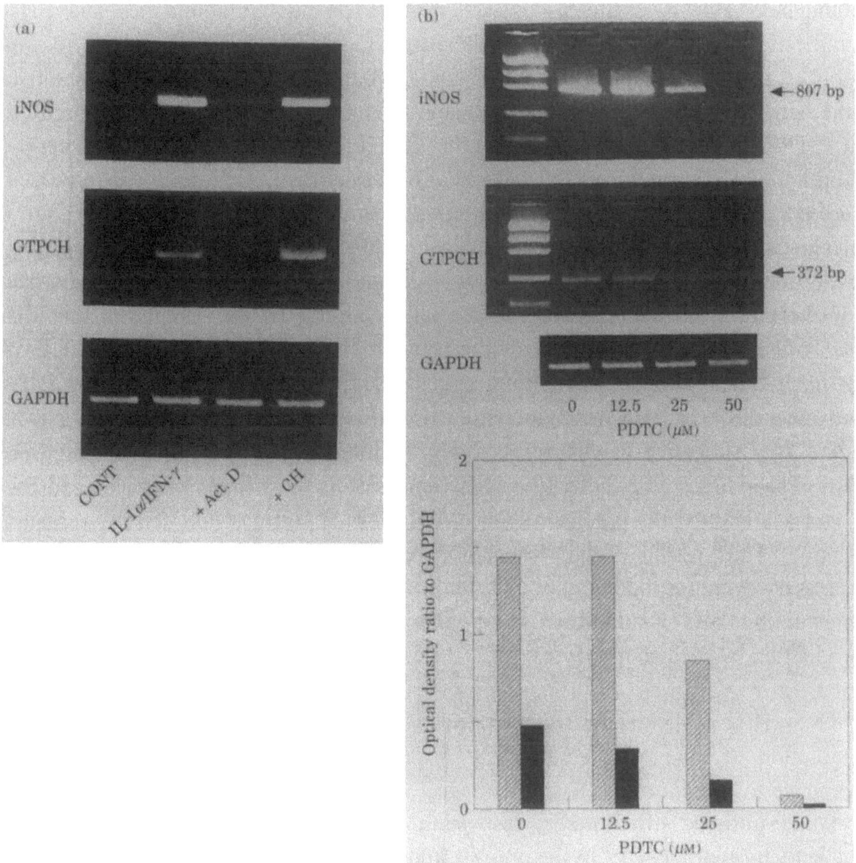

Fig. 5. (a) Effects of actinomycin D and cycloheximide on the induction of iNOS and GTPCH mRNAs. Rat cardiac myocytes were untreated (CONT), or incubated for 8 h with IL-/IFN alone (IL-1/IFN), or in the presence of actinomycin D (Act. D: 0.5 μg/ml) or cycloheximide (CH: 1 μg/ml). (b) Effects of PDTC on the induction of iNOS and GTPCH mRNA. Rat cardiac myocytes were incubated for 8 h with IL-1/IFN in the presence of the indicated concntrations of PDTC. After incubation, RNA was prepared and subjected to RT-PCR with primers specific for iNOS, GTPCH, and GAPDH as indicated. In (b), arrows indicate the predicted size of PCR products, and DNA size markers in the left lanes correspond to: 2000, 1200, 800, 400, and 200 bp. Bar graph ((b) lower pannel) is showing the relative amounts of iNOS and GTPCH mRNAs quantified by densitometry and expressed as optical density ratio of iNOS or GTPCH to GAPDH. Hatched bars: iNOS/GAPDH, black bars: GTPCH/GAPDH. Data are means of two different experiments that yielded similar results.

experiments descibed here, we have determined that specific recombinant cytokines added to primary cultures of cardiac myocytes leads to an increase in iNOS mRNA abundance and NO synthesis as well as GTPCH mRNA abundance and BH4 synthesis. Our results show that synthesis of NO and BH4 in cardiac myocytes are induced in parallel. Selective pharmacological antagonists were used to establish whether NO synthesis is affected by changes in intracellular BH4 levels. Inhibition of GTPCH or sepiapterin reductase by DAHP or NAS, respectively, potently inhibited IL-1/IFN-induced NO synthesis. Since neither DAHP nor NAS decreased cardiac myocyte respiration, nonspecific cytotoxicity was eliminated as a mechanism for inhibition of the induced NO synthesis (data not shown). Moreover, inhibition of NO synthesis by DAHP or NAS was circumvented by SEP, a substrate for the pterin salvage pathway. This latter finding strongly suggests that the mechanism for inhibition of NO synthesis by DAHP or NAS is specific blockade of *de novo* BH4 synthesis. Providing excess BH4, via administration of the BH4 precursor SEP, or BH4 itself, increased IL-1/IFN-induced NO synthesis to 200% and 170% of control values, respectively. While the mechanism of potentiation IL-1/IFN-induced NOS activity by SEP and BH4 is most likely via enhanced co-factor availability, the precise role(s) of BH4 for supporting NOS activity remains elusive. Thus, results of the present study clearly indicate that BH4 synthesis is a necessary event for induction of NO synthesis in cardiac myocytes, and conversely, interrupting BH4 availability is a potentially useful strategy for limitimg NO production by cardiac myocytes.

NF-κB is a pleiotropic transcription factor which plays a critical role in the regulation of many immediate-early response genes associated with the host response to infection and tissue damage. Therefore, it is not surprising that the cytokine-initiated inflammatory respose in cardiac myocytes may be mediated by the activation of NF-κB. In the present study, we have shown that PDTC, which is known to prevent NF-κB activation possibly via antioxidant mechanism [32], inhibits GTPCH gene expression and BH4 synthesis as well as iNOS gene expression and NO synthesis in IL-1/IFN-stimulated rat cardiac myocytes. Another inhibitor of NF-κB activation *tosyl*-lysine-chloromethyl ketone, a protease inhibitor, which inhibits NF-κB activation by inhibiting IkB-α proteolysis [33], also inhibits NO synthesis induced by IL-1/IFN in cardiac myocytes (datat not shown). These findings suggest that NF-κB plays an important role in the regulation of gene expression of GTPCH as well as iNOS in cytokine-activated cardiac myocytes.

Our study does not identify the site at which NF-κB participates in the signal transduction cascade between IL-1/IFN stimulation of cardiac myocytes and induction of iNOS or GTPCH. A protein synthesis inhibitor cycloheximide did not change the IL-1-induced mRNA abundance for iNOS and rather increased that for GTPCH, indicating no requirement for intermediary protein synthesis in the process of gene expression for these two enzyme. In addition to iNOS, cloning of the 5'-

flanking region of the rat GTPCH gene revealed the presence of several NF-κB-binding sites although these have not yet been shown to control gene expression [34]. Taking together, PDTC appeared to inhibit NO and BH4 synthesis by preventing the expression of iNOS and GTPCH genes. Thus, NF-κB may participate in the regulation of iNOS and GTPCH expression at the transcriptional level.

We conclude that immunostimulants coinduce iNOS and GTPCH gene expression; both events are necessary for activation of cellular NO synthesis in cardiac myocytes and are regulated, at least in part, by the common mechanism that is NF-κB dependent. The possible contribution of excess NO production to the decline in cardiac function raises the possibility of novel therapeutic targets and strategies for the treatment of immune-activated heart diseases.

References

1. Barry WH (1994) Mechanisms of immune-mediated myocyte injury. Circulation 89: 2421-2432
2. Lange LG, Schreiner GF (1994) Immune mechanisms of cardiac disease. N Engl J Med 174: 493-496
3. Dallman MJ, Larsen CP, Morris PJ (1991) Cytokine gene transcription in vascularized organ grafts: analysis using semiquantitative polymerase chain reaction. J Exp Med 174: 493-496
4. Balligand J-L, Kelley RA, Marsden, PA, Smith TW, Mitchel T (1993) Control of cardiac muscle cell function by an endogenous nitric oxide signalling system. Proc Natl Acad Sci USA 90: 347-351
5. Balligand J-L, Ungreanu D, Kelley RA, Kobzik L, Pimental D, Michel T, Smith TW (1993) Abnormal contractile function due to induction of nitric oxide synthesis in rat cardiac myocytes follows exposure to activated macrophage-conditioned medium. J Clin Invest 91: 2314-2319
6. Schulz R, Panas DL, Catena R, Moncada S., Olley PM, Lopaschuk GL (1995) The role of nitric oxide in cardiac depression induced by interleukin-1b and tumour necrosis factor-a. Brit J Pharmacol 114: 27-34
7. Nathan C, Xie Q-W (1994) Regulation of biosynthesis of nitric oxide. J Biol Chem 269: 25722-25729
8. Werner-Felmayer G, Werner ER, Fuchs D, Hausen A., Reibnegger G & Wachter H (1990) Tetrahydrobiopterin-dependent formation of nitrite and nitrate in murine fibroblasts. J Exp Med 172: 1599-1607
9. Gross SS, Jaffe EA, Levi R, Kilbourn RG (1991) Cytokine-activated endothelial cells express an isotype of nitric oxide synthase which is tetrahydrobiopterin-dependent

calmodullin-independent and inhibited by arginine analogs with a rank-order of potency characteristic of activated macrophages. Biochem Biophys Res Commun 178: 823-829
10. Gross SS, Levi R (1992) Tetrahydrobiopterin synthesis: An absolute requirement for cytokine-induced nitric oxide generation by vascular smooth muscle. J Biol Chem 267: 25722-25729
11. Sakai N, Kaufmann S, Milstien S (1993) Tetrahydrobiopterin is required for cytokine-induced nitric oxide production in a murine macrophage cell line (RAW264). Mol Pharmacol 43: 6-10
12. Sakai N, Kaufmann S, Milstien S (1995) Parallel induction of nitric oxide and tetrahydrobiopterin synthesis by cytokines in rat glia cells. J Neurochem 65: 895-902
13. Kasai K, Hattori Y, Nakanishi N, Manaka K, Banaba N, Motohashi S, Shimoda S (1995) Regulation of inducible nitric oxide production by cytokines in human thyrocytes in culture. Endocrinology 136: 4261-4270
14. Werner-Felmayer G, Werner ER, Fuchs D, Hausen A., Reibnegger G Schmidt K, Weiss G, Wachter H (1993) Pterindine biosynthesis in human endothelial cells: Impact on nitric oxide-mediated formation of cyclic GMP. J Biol Chem 268: 1842-1846
15. Rosenkranz-Weiss P, Sessa WC, Milstien S, Kaufman S, Watson CA, Pober JS (1994) Regulation of nitric oxide synthesis by proinflammatory cytokines in human umbilical vein endothelial cells. J Clin Invest 93: 2236-2243
16. Cosentino F, Katusic ZS. Tetrahydrobiopterin and dysfunction of endothelial nitric oxide synthase in coronary arteries. Circulation 1995;91:139-144
17. Nathan C, Xie QW, 1994. Regulation of biosynthesis of nitric oxide. J Biol Chem 269: 25722-25729
18. Xie QW, Kashiwabara Y, Nathan C (1994) Role of transcription factor NF-kB/Rel in induction of nitric oxide synthase. J Biol Chem 269: 4705-4708
19. Nunokawa Y, Oikawa S, Tanaka S (1996) Human inducible nitric oxide synthase gene is transcriptionally regulated by nuclear factor-kB dependent mechanism. Biochem Biophys Res Commun 223: 347-352
20. Mulsch A, Schray-Utz B, Mordvibtcev PI, Hauschildt S, Busse R (1993) Diethyldithiocarbamate inhibits induction of macrophage NO synthase. FEBS lett 321: 215-218
21. Sherman MP, Aeberhard EE, Wong VZ, Griscavage JM, Ignarro LJ, 1993. Pyrrolidine dithiocarbamate inhibits induction of nitric oxide synthase activity in rat alveolar macrophages, Biochem Biophys Res Commun 191: 1301-1308
22. Eberhardt W, Kunz D, Pfeilschifer J (1994) Pyrrolidine dithiocarbamate differentially affects interleukin 1b- and cAMP- induced nitric oxide synthase expression in rat mesangial cells. Biochem Biophys Res Commun 200: 163-170
23. Nichol CA., Smith GK, Duch DS (1985) Biosynthesis and metabolism of tetrahydrobiopterin and molybdopterin. Annu Rev Biochem 54:729-764
24. Hattori Y, Gross SS (1993) GTP cyclohudrolase I mRNA is induced by LPS in vascular smooth muscle: charaterization, sequence and relationship to nitric oxide synthase. Biochem Biophys Res Commun 195: 435-441
25. Gross SS, Levi R (1992) Tetrahydrobiopterin synthesis: an absolute requirement for cytokine-induced nitric oxide generation by vascular smooth muscle. J Biol Chem 267: 25722-25729
26. Hattori Y, So S, Hattori S, Kasai K, Shimoda S (1995) Vesnarinone inhibits induction of nitric oxide synthase in J774 macrophages and rat cardiac myocytes in culture. Cardiovasc Res 30: 187-192

27. Kamitani T, Ikeda U, Muto S, Kawakami K, Nagano K, Tsuruya Y, Oguchi A, Yamamoto K, Hara Y, Kojima T, Medford RM, Shimada K (1992) Regulation of Na, K-ATPase gene expression by thyroid hormone in rat cardiocytes. Circ Res 71: 1457-1464
28. Mosmann T (1983) Rapid colorimetric assay for cellular growth and survival: application to proliferation and cytotoxicity assays. J Immunol Methods 655: 55-63
29. Fukushima T, Nixon JC (1980) Analysis of reduced forms of biopterin in biological tissues and fluids. Anal Biochem 102: 176-188
30. Lowry OH, Rosebrough NJ, Farr AL, Randall RJ (1951) Protein measurement with the Folin phenol reagent. J Biol Chem 193: 265-275
31. Terada Y, Tomita K, Nonoguchi H, Marumo F (1992) Polymerase chain reaction localization of constitutive nitric oxide synthase and soluble guanylate cyclase messenger RNAs in microdissected rat nephron segments. J Clin Invest 90: 659-665
32. Schreck R, Meier B, Mannel DN, Droge W, Baeueule PA (1992) Dithiocarbamates as potent inhibitors of nuclear factor kB activation in intact cells. J Exp Med 175: 1181-1194
33. Chenn CC, Rosenbloom CL, Anderson DC, Manning AM (1995) Selective inhibition of E-selectin, vascular cell adhesion molecule-1, and intercellular adhesion molecule-1 expression by inhibitors of IkB-a phosphorylation. J Immunol 155: 3538-3545
34. Smith J, Gross SS (1996) Cloning and regulation of the rat GTP cyclohydrolase gene: co-ordinate transcriptional induction with iNOS by immunostimulants. In: Moncada S, Stamler J, Gross SS, Higgs EA (eds), The Biology of Nitric Oxide Part5, Portland Press, London, pp156

Part 4

NO and Liver

Regulation and Function of Nitric Oxide in the Liver

Bradley S. Taylor, M.D., Timothy R. Billiar, M.D., David A. Geller, M.D.[1]

SUMMARY. The inducible nitric oxide synthase (iNOS) gene is expressed in nearly every organ and cell type during endotoxemia. Previously, we have shown that a combination of cytokines synergistically activate iNOS expression in the liver, and we have cloned the first human iNOS gene from cytokine-stimulated hepatocytes. We have also shown that steriods, TGF-β, the heat shock response. and nitric oxide (NO) itself, all down-regulate iNOS expression. In vivo, we have shown that hepatic iNOS induction is differentially regulated from the typical acute-phase reactants and is not expressed as a mandatory component of the acute phase response. Thus, numerous mechanisms have evolved to regulate iNOS expression during hepatocellular injury.

Studies into the function of NO in the liver by our group have shown that induced NO synthesis plays a protective role in the liver during septic and inflammatory conditions. In a rodent model of endotoxemia, the addition of the non-specific NOS inhibitors significantly increased hepatic damage. NO exerted a protective effect through its ability to prevent intravascular thrombosis by inhibiting platelet adhesion and neutralizing toxic oxygen radicals. Recently, we have shown that NO exerts protective effects both in vivo and in vitro by blocking TNFα-induced apoptosis and hepatotoxicity, in part by a thiol-dependent inhibition of caspase-3-like protease activity. Taken together, these studies demonstrate the cytoprotective effects of NO in the liver and suggest that hepatic iNOS expression functions as an adaptive response to minimize inflammatory injury.

KEY WORDS: Nitric oxide, Inducible nitric oxide synthase, Gene regulation, Apoptosis

[1] Department of Surgery, University of Pittsburgh, Pittsburgh, PA 15261, USA

Introduction

Nitric oxide (NO), the end product of the enzymes nitric oxide synthase (NOS), is a potent biologic mediator with diverse physiologic and pathophysiologic properties. NO functions to regulate blood pressure, neurotransmission, antimicrobial defense mechanisms, and modulate the inflammatory response [1]. Once produced, NO has a short half life (t1/2=seconds) and undergoes spontaneous oxidation to the inactive metabolites nitrite and nitrate (NO_2^- and NO_3^-). This chapter reviews the regulation and function of the inducible NO synthase (iNOS) isoform in the liver. The regulation of iNOS is complex and occurs at multiple levels in the pathway of gene expression. Also, the roles of NO in regulating hepatic function and disease will be discussed.

Cloning the inducible nitric oxide synthase gene

Our group isolated the first human iNOS cDNA from LPS- and cytokine-stimulated primary human hepatocytes in 1993 [2]. The sequence of the human hepatocyte iNOS clone reveals a 4,145 base pair cDNA containing a 3,459 base pair open reading frame which encodes a polypeptide of 1,153 amino acids with a calculated molecular mass of 131 kDa. Compared to human endothelial cNOS [3] and human neuronal cNOS [4], human hepatocyte iNOS displays 51 % and 53 % amino acid sequence identity, respectively. Overall, human iNOS displays a 80 % sequence identity to murine macrophage iNOS at both the nucleotide and amino acid levels. The hepatocyte iNOS sequence is 6 amino acids longer in the amino portion of the protein and 3 amino acids longer at the carboxyl terminus as compared to the murine macrophage iNOS. Similar to other NOS isoforms, hepatocyte iNOS contains consensus recognition sites for the cofactors FMN, FAD, and NADPH in the caboxyl half of the protein which have been shown to be important for iNOS enzyme activity. In addition, a consensus sequence for a calmodulin recognition site is also present. The functional role of the human iNOS cDNA was confirmed by transfection of an iNOS cDNA expression vector (Genetech) into human kidney cells, which resulted in a substantial increase in NOS activity as detected by conversion of radiolabeled arginine to citrulline [2]. Since our initial report, human iNOS cDNAs have been cloned from chondrocytes [5], DLD-1 colon carcinoma cell line [6], and A-172 glioblastoma cell line [7]. Each of these cDNAs are identical to the human hepatocyte iNOS cDNA with >99% sequence homology.

Next, using the iNOS cDNA to screen a fibroblast genomic library, the human iNOS gene was isolated and subsequently cloned [8]. Briefly, two rounds of screening yielded 7 distinct cosmid clones which were aligned by restriction mapping, Southern blot hybridization, and DNA sequencing analysis. All of the isolated clones were found to be part of a single genomic locus. The full-length human iNOS gene is 37 kb in length and is composed of 26 exons and 25 introns. This genomic structure is similar to that of human endothelial and neuronal NOS genes, and suggests the divergence from a common ancestor. Primer extension analysis of LPS and cytokine-stimulated human hepatocyte RNA identified the transcriptional initiation site 30 base pairs downstream of the TATA box. By utilizing a somatic cell hybrid mapping panel and flourescent *in situ* hybridization, the human iNOS gene was mapped to chromosome 17 at position 17 cen q11.2 [8]. The human neuronal and endothelial NOS reside on chromosome 12 and 7, respectively, confirming that the three NOS genes are distinct.

Molecular regulation of nitric oxide synthase in the river

The cDNA probe was then used to study the induction of iNOS mRNA in human hepatocytes. Northern blot analysis was performed on RNA from cultured human hepatocytes stimulated with LPS, TNFα, IL-1β, and IFNγ for 2-48 hours. No iNOS mRNA was detected in the unstimulated human hepatocytes [9]. However, a single mRNA band at ~4.5 kb first appeared 4 hours after cytokine and LPS stimulation, peaked at 8 hours, and was barely detectable by 48 hours. To correlate mRNA levels with iNOS activity, NO_2^- and NO_3^- levels were measured from the culture supernatants and were found to increase 20- to 30-fold at 24 and 48 hours after stimulation. While most cell types require a combination of cytokines to activate iNOS expression, IL-1β alone at high doses can induce iNOS mRNA in cultures of primary human hepatocytes [10] and chondrocytes [5]. Because primary human hepatocytes do not undergo mitosis in culture, a human liver epithelial cell line (designated AKN-1) was isolated. These cells are derived from normal human liver and immunohistochemical staining and karotype analysis revealed a transformed liver cell line with both biliary and epithelial cell features. When stimulated with the cytokine mix (CM) of TNFα, IL-1β, and INF-γ, iNOS mRNA was observed at 2 hours, peaked at 4 hours, and diminished by 24 hours (Fig. 1). NO release lagged behind the iNOS mRNA expression in a time dependent fashion. Stimulation with single cytokines revealed that only IFNγ weakly stimulated AKN-1 mRNA expression. Double and triple cytokine combinations had a synergistic effect with CM inducing maximal iNOS mRNA expression. These results are similar to those seen in primary human hepatocytes and indicate that the AKN-1 cell line behave similarly to primary human hepatocytes and thus would be useful for studying the regulation of human iNOS gene expression. We then tested the hypothesis that the human iNOS gene is regulated in part by gene transcription. Nuclear run-on analysis in the AKN-1 cells showed a 5-fold increase in cytokine induced transcriptional activity of iNOS [11]. This up-regulation appears to be mediated by the transcription factor NF-κB in both human (unpublished data) and rat hepatocytes [12,13]. Interestingly, however, the amount of NO produced by human cells is lower than that seen with their rodent counterparts which we have used extensively in our studies of iNOS regulation.

Besides human cells, the iNOS cDNA has been cloned from both mouse and rat cells and has been a useful research tool for studying the molecular regulation of

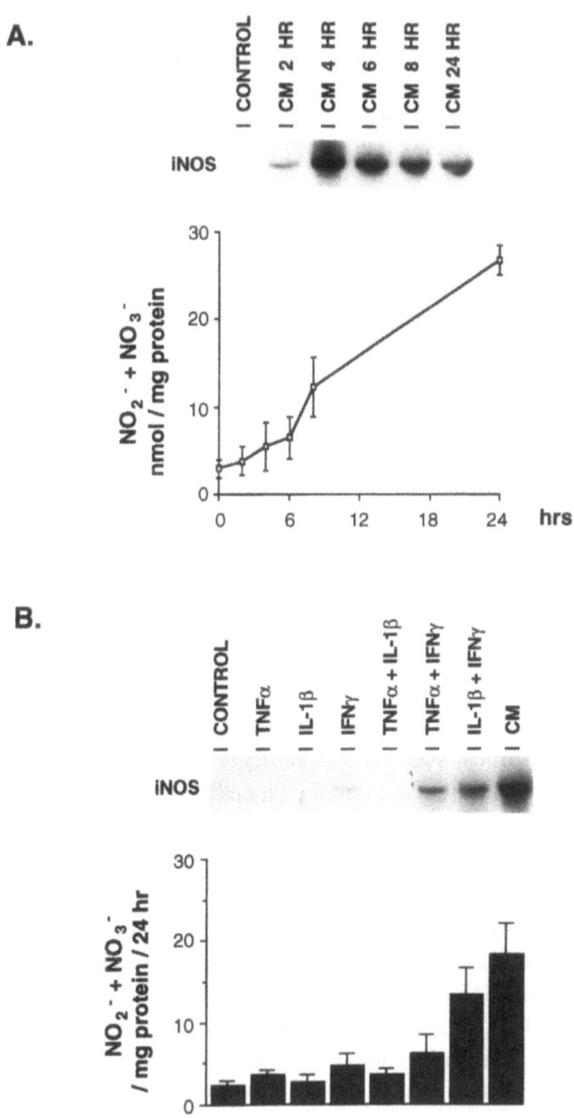

Fig. 1. Cytokine-induced iNOS expression in AKN-1 cells A) Time course of iNOS mRNA induction (Northern blot, upper) and nitrite and nitrate (NO_2^- and NO_3^-) production in culture supernatants (lower) following stimulation with TNFα (1000 u/ml) + IL-1β (100 u/ml) + IFNγ (250 u/ml) (cytokine mix, CM). B) Pattern of 6-hr. iNOS mRNA induction (Northern blot, upper) and NO_2^- and NO_3^- release (lower) in response to cytokines and cytokine combinations. Values for NO_2^- and NO_3^- are expressed as mean ± SEM for three independent experiments. (cytokine mix, CM).

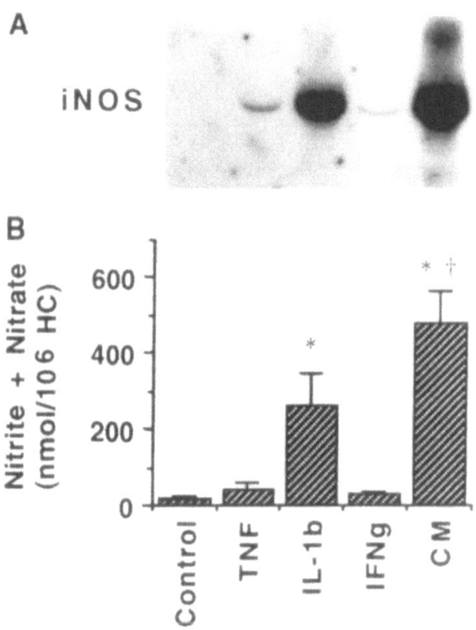

Fig. 2. The effect of different cytokines on induced hepatocyte NO synthesis. Cultured rat hepatocytes were stimulated with cytokines TNFα (500 u/ml), IL-1β (500 u/ml), or IFNγ (500 u/ml), or the CM of TNFα + IL-1β+ IFNγ (each at 500 u/ml). (A) Northern blot ana- lysis for iNOS mRNA 6 h after stimulation, as indicated in the columns below. (B) Hepatocyte NO synthesis was determined by measuring NO_2^- and NO_3^- levels in the culture supernatants. Values are expressed as nmoles (nmol) NO_2^- and $NO_3^-/10^6$ hepatocytes (HC)/24 h and represent the mean ± SE (n=6 per group). * indicates p<0.01 vs control, p<0.01 vs. IL-1β (by ANOVA). † indicates p <0.01 vs IL-1β.

iNOS. Rodent macrophages were among the first cell types that were found capable of producing NO from iNOS. In response to LPS with or without IFN-γ, iNOS mRNA and protein are produced, and large amounts of NO_2^- and NO_3^- are detected [14]. Rat Kupffer cells are similar to other macrophages in their ability to respond to inflammatory stimuli with iNOS induction [15]. It was later demonstrated that hepatocytes produce NO [16]; this was the first evidence that a parenchymal cell type could express iNOS. Hepatocytes readily express iNOS in response to conditioned medium from LPS and IFN-γ-stimulated Kupffer cells [16]. It has been shown that the Kupffer cell secretory products responsible for the induction of iNOS in hepatocytes are TNFα and IL-1β [17], and it appears that IL-1β is the more important stimulus (Fig. 2) [10]. These Kupffer cell-derived cytokines act synergistically with LPS and IFN-γ to substantially induce iNOS expression in hepatocytes [17,18]. In LPS-primed hepatocytes [19], or rats injected with turpentine to produce remote tissue injury [20], single cytokines such as TNF, IL-1 or even LPS alone are sufficient to induce NO synthesis. In experimental models of *in vivo* endotoxemia, hepatocytes readily express iNOS mRNA, which can be detected by Northern analysis within 3 hours, and peak in iNOS enzyme

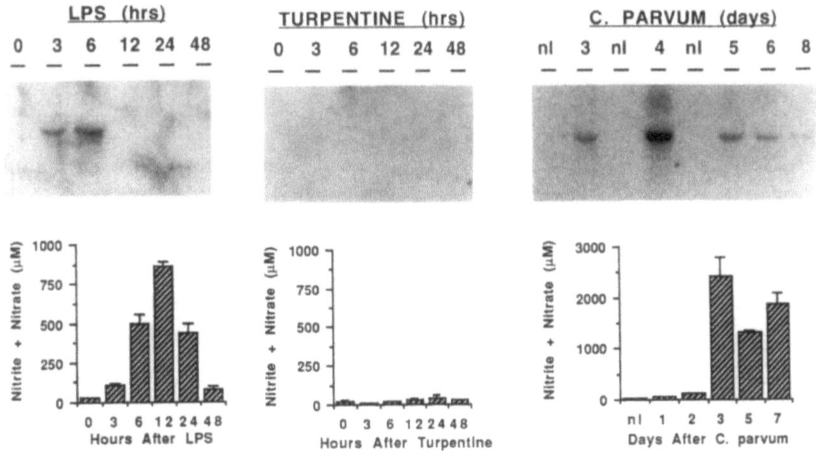

Fig. 3. Northerrn blot analysis and NO release data from animals treated with either LPS, turpentine, and C. Parvum injection.

levels at 12-16 hours. Plasma NO_2^- and NO_3^- subsequently peaked and then declined rapidly [21].

The iNOS gene is differentially regulated in vivo and is not expressed as a mandatory component of the hepatic acute phase response. Hepatocytes express large amounts of iNOS in response to LPS and killed *Corynebacterium parvum* (*C. parvum*) injection, a known inducer IL-1, TNF-α, and IFN-γ (Fig. 3) [21]. However, in contrast to LPS and *C. Parvum* injection, hindlimb intramuscular injection of turpentine resulted in local tissue destruction, marked induction of the hepatic acute phase response without inducing iNOS . This demonstrates that cytokine regulation of iNOS is different from that of the well characterized acute phase response.

Besides defining what factors upregulated iNOS gene expression, our laboratory is interested in defining those factors which down-regulate iNOS expression. Glucocorticoids are known to inhibit induced NO synthesis in several cell types [22]. We found that dexamethasone decreases iNOS mRNA levels in rat hepatocytes and that this down-regulation occurs at the level of transcription (Fig. 4)[23]. This effect is a result of increased cytosolic I-kBα levels and a concomitant decrease in nuclear p65 translocation in the presence of dexamethasone [13]. Interestingly, this effect appears to be tissue and species specific. For example, Kleinert et al suggest that dexamethasone inhibits cytokine-induced iNOS mRNA in human A549 epithelial cells by a direct protein-protein interaction between the glucocorticoid receptor and NF-κB without an increase in IκBα mRNA levels [24].

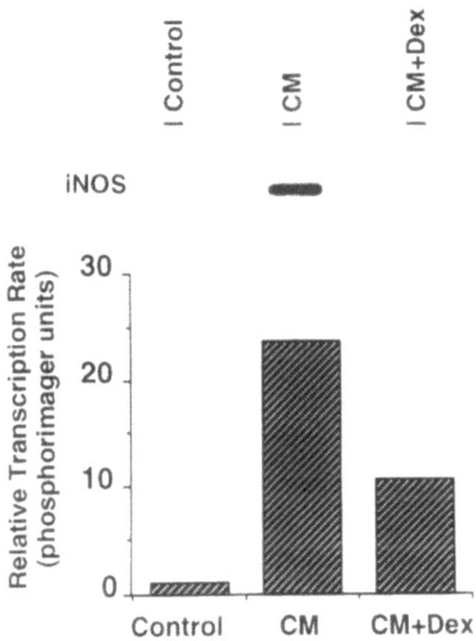

Fig. 4. Nuclear run-on analysis of cytokine-induced iNOS message in AKN-1 cells. Control cells were not exposed to cytokines. Cytokine-stimulation for 2 and 4 hours was conducted with TNFα (1000 u/ml) + IL-1β (100 u/ml) + IFNγ (250 u/ml) (cytokine mix, CM). Nuclei were isolated and incubated with radiolabelled CTP. ^{32}P-labelled nuclear RNA was hybridized to a GeneScreen membrane containing 5μg of immobilized sense (S) and antisense (AS) cRNA probes of iNOS, glutaminase (GLN), and argininosuccinate synthetase (AGS).

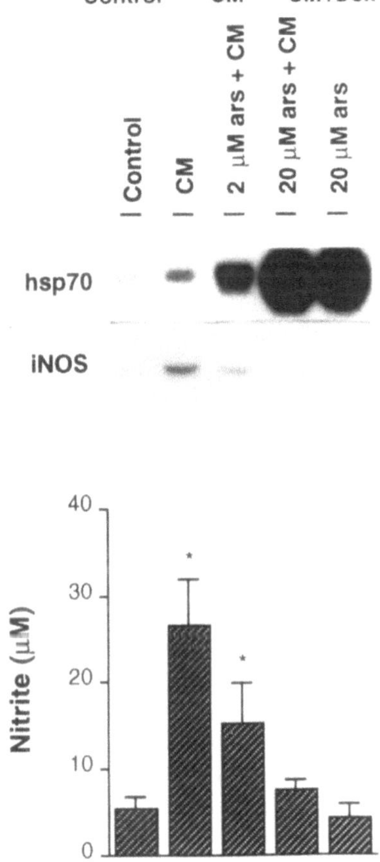

Fig. 5. Effect of prior induction of the HSR on iNOS gene expression and NO synthesis. Following pre-induction of the HSR by Ars (2 or 20 mmol/L), cultured rat hepatocytes were treated with cytokine mix of TNFα (500 U/ml) + IL-1β (200 U/ml) + IFNγ (100 U/ml) to stimulate iNOS expression. Northern blot analysis for hsp 70 and iNOS mRNA was performed and 24 h NO_2^- release measured.

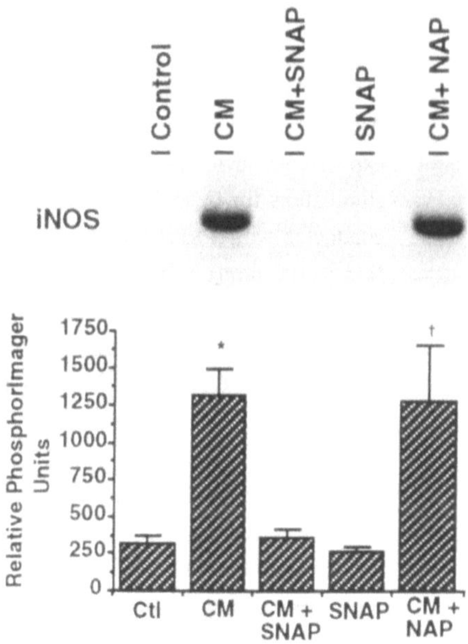

Fig. 6. Nitric oxides effect in iNOS mRNA expression in rat hepatocytes stimulated with cytokine-mix. Cultured rat hepatocytes were treated with cytokine-mix (CM) of TNFα (500 U/ml) + IL-1β (200 U/ml) + IFNγ (100 U/ml) with or without SNAP (1000 μM) or NAP (1000 μM) for 6 hours. A) Northern blot analysis for iNOS mRNA. Equal loading of RNA in each lane was confirmed by 18S ribosomal RNA probe (not shown). B) Graph indicating relative fold induction in mRNA levels by PhosphorImager analysis (n=5 per group). * indicates $p<0.01$ vs. control, † indicates $p<0.01$ vs. CM, (by ANOVA).

Additional studies have demonstrated that the induction of the heat shock response by hyperthermia or sodium arsenite exposure blocked subsequent iNOS expression in AKN-1 human liver cells (Fig. 5) [25] as well as in primary rat hepatocytes [26]. The inhibitory effect of heat shock on iNOS expression was observed in promoter transfection experiments, however there was no effect of heat shock response on iNOS enzyme activity in AKN-1 cells transduced to stably express iNOS. These results indicate that prior induction of the heat shock response inhibits iNOS gene transcription, but not iNOS translation or enzyme activity. These findings underscore the complex array of phenotypic responses in the liver during stress conditions. Similarly, Feinstein et al have shown that heat shock protein 70 mediates suppresssion of iNOS expression in brain astroglial cells by inhibiting NFκB-DNA-binding activity [27]. The induction of the heat shock response appears to be an adaptive response to prevent over-expression of iNOS during certain inflammatory conditions. This is supported by the observation that increased NO exposure stimulates the heat shock response [28].

Recently, we have shown that NO inhibits its own gene expression, at the level of transcription, by down-regulating NF-κB expression in rat hepatocytes [12] and primary human hepatocytes (unpublished data). iNOS mRNA, protein, and NO release were markedly suppressed in the presence of either endogenously produced NO or an NO donor (Fig. 6). As levels of NO increase, feedback regulation begins and NO negatively modulates NF-κB DNA binding activity and enzyme activity to tightly regulate the amount of NO produced (Fig. 7). This data identifies a novel negative feedback loop whereby NO down-regulates iNOS gene expression.

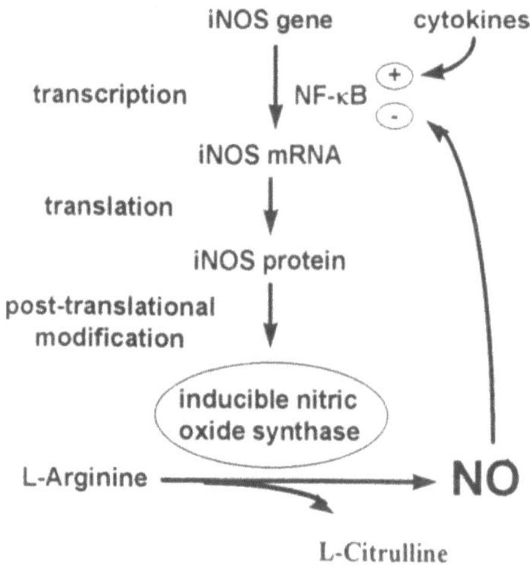

Fig. 7. Proposed mechanism of negative feedback inhibition of iNOS gene expression by nitric oxide. Cytokines stimulation activates NF-κB nuclear translocation and DNA-binding to the iNOS gene promoter resulting in iNOS gene transcription. iNOS mRNA is then translated to iNOS protein allowing for induced NO synthesis from L-arginine. Upon increasing levels of nitric oxide, feedback regulation occurs whereby, nitric oxide inhibits iNOS gene transcription, and to a lesser extent enzyme activity.

possibly to limit over production during pathophysiologic conditions. By understanding the molecular regulation of iNOS expression, therapeutic modalities can be designed to govern its induction. The nuclear transcription factor NF-κB is one potential target for such anti-inflammatory therapy.

NO has been shown to cause DNA strand breaks as well as mutations in human cells. Since p53 plays an important role in cellular response to DNA damage from exogenous mutagens, the hypothesis that p53 may perform a similar role in regulating iNOS expression was tested. Exposure of human cells to a NO donor resulted in p53 accumulation. Also, expression of p53 in multiple human cell lines, including the AKN-1 liver cell line, resulted in a down-regulation of iNOS gene expression by inhibiting the iNOS promoter [29]. These data imply a negative feedback loop where NO-induced DNA damage activates p53 expression which then mediates repression of the human iNOS gene.

In other studies, the effects of hepatocellular mitogens on cytokine-induced iNOS expression in cultures of primary human hepatocytes were examined. Hepatocyte growth factor (HGF), epidermal growth factor (EGF), and transforming growth factor TGF-β all down-regulated LPS and cytokine-induced human iNOS mRNA, iNOS enzyme activity, and NO release [30]. Interestingly, the cytokine-stimulated NO release caused a decrease in hepatocyte protein and DNA synthesis, and this effect was partially reversed by the liver growth factors. These mitogens are known to directly trigger hepatic DNA and protein synthesis; however, our results raise the possibitity that these hepatic growth factors may also promote protein and DNA synthesis during inflammatory conditions, in part, by suppressing iNOS expression. The precise mechanism by which these growth factors exert their effects on iNOS expression is unknown.

In addition to transcriptional control, downstream regulation may also be important in the regulation of iNOS expression in the liver. The 3'-UTR of the human hepatocyte iNOS cDNA contains several AT-rich sequences which correspond to the AU sequences in the mRNA. These ATTTA sequence motifs have been shown in many labile cytokine and proto-oncogene transcripts and have been shown to destabilize mRNA [31,32] and inhibit translational effiency [33]. In addition, mechanisms of post-translational modification including protein stabilization, dimerization, phosphorylation, subcellular localization, cofactor binding and substrate availabilty play an important role in iNOS expression.

For example, experiments have also been conducted to examine the rate-limiting aspect of BH_4 availability in perfused livers and isolated hepatocytes [34]. Suppression of BH_4 levels markedly reduced the conversion of phenylalanine to tyrosine by phenylalanine hydroxylase but had little impact on iNOS activity. Furthermore, phenylalanine increased BH_4 synthesis as previously described [35] while arginine had no effect. These findings are not entirely unexpected because the iNOS Km for BH_4 is 100-fold lower than the phenylalanine hydroxylase Km for BH_4. Thus, the requirement of iNOS for BH_4 is much lower. While phenylalanine hydroxylase relies on allosteric upregulation of BH_4 synthesis by the substrate, iNOS does not.

Nitric oxide and hepatocyte function

NO is an important regulator of hepatocellular functions including protein synthesis and the metabolism of carbohydrates and drugs (Table 1). When hepatocytes are cocultured with Kupffer cells and the system is exposed to LPS, a significant reduction in the incorporation of [^{3}H]leucine occurs, signifying an inhibition of hepatocyte protein synthesis [36]. This inhibition of protein synthesis is accompanied by increased levels of NO_2^- and NO_3^- in the supernatant, and can be prevented with L-NMMA [37], confirming that NO produced by Kupffer cells and hepatocytes can act to reduce the total protein production of hepatocytes. The addition of exogenous NO or NO donors reproduces the NO-mediated inhibition of protein synthesis [38]. Likewise, when isolated hepatocytes in primary culture are exposed to the appropriate cytokine mix, a similar decrease in protein synthesis occurs, associated with elevation in NO_2^- and NO_3^- levels. These changes are also inhibited with L-NMMA, demonstrating that endogenous hepatocyte NO synthesis can act in an autocrine fashion to inhibit protein synthesis *in vitro* [17]. In the coculture system and in primary hepatocytes in isolated culture, the production of NO appears to be the result of iNOS because there is a delay of several hours between stimulation and the release of NO [39].

The mechanism of NO-mediated suppression of hepatocyte protein synthesis is unclear. However, it appears to be cGMP-independent because cGMP analogs are

Table 1. NO-mediated effects on hepatocytes

Inhibition of protein synthesis
Inhibition of gluconeogenesis
Inhibition of glycogenolysis
Inhibition of glyceraldehyde-3-phosphate dehydrogenase
Activation of soluble guanylate cyclase
Inhibition of cytochrome P-450 1A1 and 1A2
Inhibition of mitochondrial respiration
inhibition of aconitase

unable to reproduce this effect. Nor does not involve loss of hepatocyte viability, because it is reversible with NOS inhibitors, the hepatocytes exclude trypan blue, and no increase in hepatocellular enzyme release is detected [38].

NO-mediated suppression of hepatocyte protein synthesis does not appear to occur *in vivo*. Sax et al, in a sepsis-induced model of iNOS expression demonstrated increased hepatocyte protein synthesis [40]. This same group went on to document a decrease in the level of in vivo protein synthesis during endotoxemia. Whereas Frederick et al in a cecal ligation/puncture model found a decrease in the level of *in vivo* hepatic protein synthesis with administration of NOS inhibitors [41].

The liver is an important organ in the metabolism of carbohydrates. Liver glycogen metabolism is controlled by hormones such as glucagon and epinephrine. *In vitro* data suggests that NO mediates an inhibitory effect on carbohydrate metabolism in hepatocytes. Brass and coworkers showed that the NO donor *S*-Nitroso-*N*-acetyl-penicillamine (SNAP) inhibits cyclic adenosine monophosphate (cAMP) and glucagon-stimulated hepatic glycogenolysis [42], suggesting that glucose homeostasis may be partially controlled by NO. Furthermore, Horton and coworkers demonstrated that NO donors inhibit cultured hepatocyte gluconeogenesis [43] raising the possibility that NO could account for the suppressed gluconeogenesis encountered in models of sepsis. In other *in vitro* work, we and others have demonstrated a marked NO-dependent inhibition of the activity of glyceraldehyde-3-phosphate dehydrogenase in rat livers with high-level iNOS expression [44]. However, we were unable to demonstrate an NO-dependent suppression of gluconeogenesis in LPS-treated rats using the NOS inhibitor L-NMMA (non-specific NOS inhibitor) or aminoguanidine (iNOS-specific inhibitor). Similarly, no difference in the LPS-mediated decrease in hepatic gluconeogenesis was seen between iNOS knockout and iNOS wild-type mice, making the role of NO in suppression of gluconeogenesis questionable [45]. Thus, while NO has been shown to inhibit hepatic gluconeogenesis in cell culture, there is to date no evidence for this *in vivo*. The differences between hepatocyte protein synthesis and carbohydrate metabolism underscores the importance of confirming *in vitro* data with *in vivo* experiments.

One of the earliest actions of NO to be defined was that NO binds avidly to heme prosthetic groups and the iron sulfur complexes in certain enzymes [46]. This interaction may result in enzyme activation or inhibition. In the liver, a number of enzymes are potential targets of this type of interaction. Like in other cell types, soluble guanylate cyclase activity in hepatocytes is regulated by NO. In response to NO-generating compounds, rat liver slices release cGMP [47], while cultured hepatocytes stimulated to express iNOS produce moderate amounts of cGMP [48]. While all of the roles of cGMP in hepatocyte physiology have not been determined, we have recently shown that NO-stimulated cGMP production inhibits TNF-α-induced apoptosis in hepatocytes [49]. High levels of NO inhibit hepatocyte mitochondrial respiration *in vitro* [50,51]. However, evidence that NO inhibits mitochondrial function *in vivo* is lacking.

The liver metabolizes a large number of drugs, toxins, carcinogens and other metabolic products via the heme-containing cytochrome P-450 class of proteins [52]. Cytochrome P-450 activity in hepatocytes is inhibited in coculture with cytokine-stimulated Kupffer cells which produce NO [53]. Furthermore, inflammatory conditions that induce iNOS in hepatocytes also result in inhibition of cytochrome P-450 activity. The administration of L-NAME prevents the inhibition of cytochrome P-450 in LPS-stimulated rats [54] and in hepatocytes stimulated to produce NO [55]. Thus, both exogenous and endogenous NO can affect the activity of this important enzyme system, and this may explain the clinical observation of altered hepatic metabolism of certain drugs and toxic substances in the setting of systemic inflammatory processes and liver dysfunction.

The cytoprotective role of no in the liver

Endotoxemia and systemic inflammation can result in liver damage as detected histologically and by the presence of liver enzymes in the plasma. In these settings iNOS is upregulated and expressed in hepatocytes and Kupffer cells [21]. The finding of liver iNOS induction in sepsis is also supported in humans. Elevated NO_2^- and NO_3^- levels are detected in the plasma of septic patients [56]. A large volume of experimental work has attempted to determine the role of NO in sepsis in the liver and in other organs, namely the vasculature, heart, lung and kidney. In the field of NO biology, both beneficial and detrimental NO-mediated properties have been described.

In various experimental models, our group has shown that nonspecific inhibition of the NOS enzyme in endotoxemia results in increased liver damage, supporting a beneficial role of NO in the liver during sepsis. In an *in vivo* murine model of endotoxemia-induced hepatic necrosis, increased NO production was hepatoprotective, while NOS inhibition resulted in markedly increased hepatic injury [57]. In this study, killed *C. parvum* injection resulted in the development of

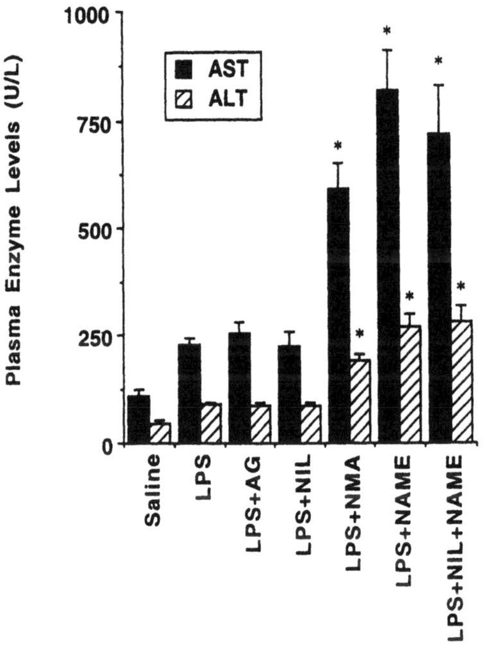

Fig. 8. NMA and NAME aggravate hepatocellular necrosis in LPS-treated animals. Rats were injected with LPS or saline following implantation with osmotic pumps for portal vein infusion of either saline or the NOS inhibitors indicated. Plasma AST and ALT levels were determined 16 hours after LPS injection. Data are the mean ± SEM from seven animals per group. (*$p<0.03$ vs saline+LPS).

hepatitis, which was then followed 5-7 days later by the administration of LPS, causing liver necrosis. L-NMMA administration resulted in increased hepatic injury, with a 3- to 5-fold increase in liver damage as detected by OCT and AST. This data suggests that NO produced locally plays a protective role in the liver during systemic inflammation. Additional studies using the same model of hepatic injury showed that co-administration of superoxide dismutase could attenuate the L-NMMA-associated exacerbation of hepatic injury [58], while cyclooxygenase inhibition worsened the injury [59]. These results suggests that reactive oxygen intermediates play a prominent role in the LPS-induced hepatic cytotoxicity. Also, prostaglandin synthesis appears to act in synergism with NO in its hepatoprotective role in the face of an LPS challenge. Histologic examination of the liver during endotoxemia with or without NOS inhibition shows increased microvascular thrombosis in mice treated with NOS inhibitors [58,60], confirming the known role of NO in maintaining vascular patency and preventing platelet aggregation and leukocyte adhesion to the endothelium. However, Szabo et al [61] showed that the selective iNOS inhibitors reduced LPS-induced damage in the rat liver. They postulated that while low-level production of NO via cNOS was cytoprotective and maintained microvascular patency, excess NO production via iNOS mediated cytotoxicity.

To further examine the role of the specific NOS isoforms we have carried out experiments where nonselective or iNOS-selective inhibitors were infused continuously into the liver of rats during endotoxemia. We found that the nonselective inhibitors (L-NMMA or L-NAME) increased both necrosis (Fig. 8) and apoptosis of hepatocytes, while the iNOS-specific inhibitors N-iminoethyl-L-lysine (L-NIL) or aminoguanidine increased only apoptosis during endotoxemia [62]. We have subsequently shown that NO has a potent anti-apoptotic effect in hepatocytes and inhibits hepatocyte apoptosis resulting from nutrient withdrawal [49]. NO appears to exert these actions by both cGMP-dependent and independent mechanisms. Therefore, we find little evidence for direct NO-mediated hepatotoxicity in endotoxemia. It appears that during sepsis, low-level NO production in the liver (via cNOS) protects against hepatocyte necrosis, while higher amounts of NO (produced by iNOS) prevent apoptosis.

The mechanisms of endotoxemia-induced hepatocyte apoptosis are beginning to be understood. In response to stimuli such as the TGF-β1 [63], Fas ligand [64] or TNF-α and D-galactosamine [65], hepatocytes will undergo the characteristic apoptotic changes such as chromosomal condensation, oligosomal DNA fragmentation and cytoplasmic blebbing. While in some cell types NO may induce apoptosis [66,67], there is a growing body of evidence that NO plays an anti-apoptotic role in the liver. *In vitro* studies aimed at determining the mechanism of this action have confirmed the previously described *in vivo* work. Fetal hepatocytes in primary culture (pre-stimulated with LPS to induce iNOS) are protected from TGF-β1-mediated apoptosis [68]. NO donors are also protected against apoptotic cell death in these cultured cells. Similarly, sodium nitroprusside confers complete protection against TNF-α-induced hepatocyte apoptosis in D-galactosamine-sensitized mice *in vivo* [69]. Several mechanisms for the anti-apoptotic effect of NO have been postulated. Our group has recently shown that the NO donor SNAP induces the expression of heat shock protein 70 (HSP70) mRNA and protein in hepatocytes, which confers protection from TNF-α and actinomycin D-induced apoptosis [28]. The degree of protection correlated directly with the level of HSP70 expression, and could be reversed by blocking HSP70 expression with an antisense oligonucleotide to HSP70. It was also suggested that causing glutathione oxidation was the mechanism by which NO induced HSP70. Another line of investigation implicates NO in inhibiting the activity of caspase-3-like protease. This pro-apoptotic enzyme family belongs to the protease signaling cascade. Hepatocytes in primary culture were stimulated to undergo apoptosis by either removal of growth factors, exposure to TNF-α or anti-Fas antibody. Inducing iNOS with CM or in the presence of exogenous NO donors substantially reduced the loss of cell viability and completely eliminated the appearance of DNA fragmentation (Figs. 9 and 10) [70]. This was associated with inhibition of caspase-3-like activity by S-nitrosylation and an indirect suppression of caspase-3-like activation via a cGMP-dependent process.

Kuo and coworkers have provided evidence that NO also protects against acetominophen-induced toxicity. In this model, inhibition of NO synthesis worsened the resultant liver injury [71]. In this study, L-NMMA administration prior to acetaminophen resulted in a doubling of the plasma aspartate aminotransferase, with a concomitant decrease in cellular glutathione (GSH) levels, suggesting that the inhibition of NO synthesis potentiates acetaminophen

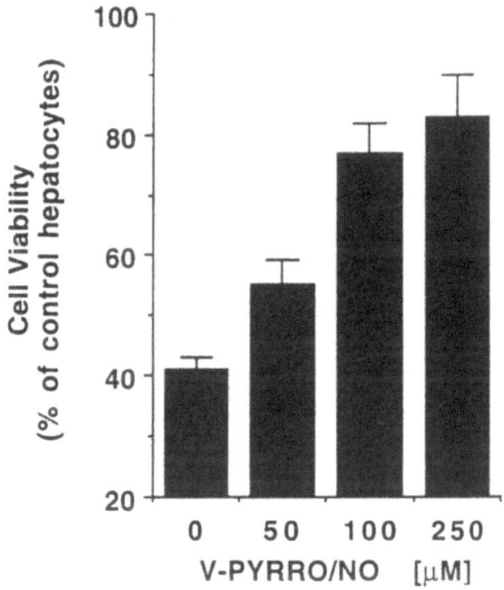

Fig. 9. Viability of cultured rat hepatocytes, as determined by crystal violet staining, at 12 h after exposure to 40 pg/ml TNFa+0.5 μg/ml ActD (TNF/ActD) in the presence of increasing concentrations of V-PYRRO/NO.

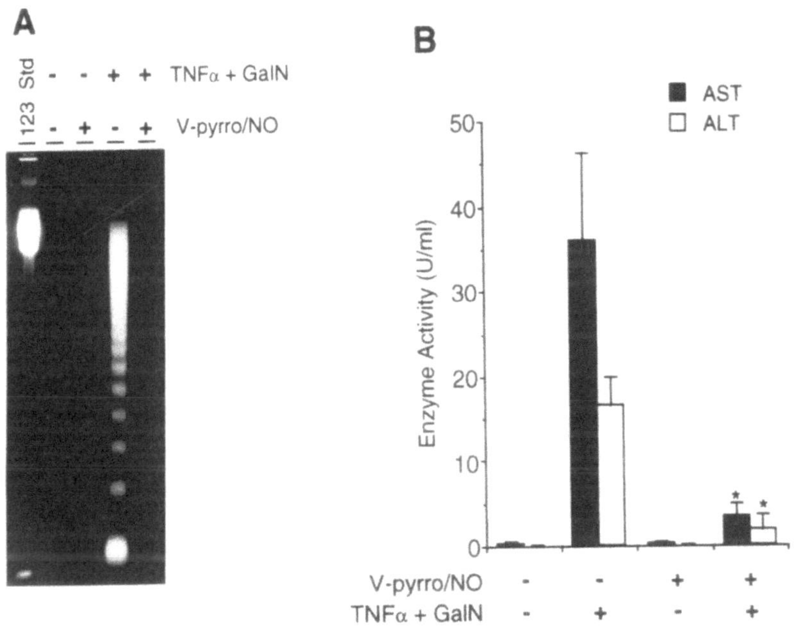

Fig. 10. Cytoprotective role of NO in vivo. (A) Hepatic apoptosis was determined 8 hours after administration of 10mg/kg TNFa+700 mg/kg galactosamine (GALN) to rats with or without constant infusion of V-PYRRO/NO (1.06 mmol/kg/h) by measuring the presence of fragmented cytosolic DNA in whole liver (data representative of six animals per group). (B) At 24 hours, plasma aspartate aminotransferase (AST) and alanine aminotransferase (ALT) levels were determined (N=5-6 per group; *P<0.01 vs. TNFa+GalN with saline infusion).

hepatotoxicity by depleting GSH stores, perhaps reducing the cellular ability to neutralize oxidant stress.

There is also evidence to suggest that NO may be beneficial in the liver under a number of pathologic conditions where reactive oxygen intermediates are produced. These substances include superoxide anion ($O_2^{\cdot}$), hydroxyl radical ($OH^{\cdot}$) and hydrogen peroxide (H_2O_2). In the liver, these reactive species can cause direct membrane and DNA damage, lipid peroxidation, and hepatocyte necrosis [72]. Under these redox conditions, NO may act as an antioxidant by reacting with toxic oxygen metabolites, producing less toxic species [73]. This may help explain the data that show a protective role for NO in murine LPS-induced hepatic necrosis, because reactive oxygen intermediates produced by LPS-stimulated macrophages are known to cause this injury [74]. Furthermore, Kim et al have shown that pretreating hepatocytes with NO renders the cells less sensitive to a subsequent exposure to H_2O_2 or high concentrations of NO [75]. NO-induced heme oxygenase, which may occur through the liberation of iron within the cell, then provides protection through the formation of biliverdin. Thus the induction of the protective mechanisms requires relatively high concentrations of NO. However, under other conditions, NO reacts with these same reactive species to produce toxic intermediates. This is the case with the production of peroxynitrite ($OONO^{\cdot}$) resulting from the reaction of NO and $O_2^{\cdot}$. Peroxynitrite can oxidize sulfhydryl groups, cause lipid peroxidation, nitration of tyrosines, and DNA base damage, and can decompose to generate other potent oxidants (76). Thus, the balance of NO protective and cytotoxic effects will depend, in part on the redox state of the cell.

Nitric oxide and liver function during ischemia/reperfusion

Liver injury secondary to ischemia-reperfusion is encountered in a number of clinical scenarios including trauma, hemorrhagic shock, liver resection and liver transplantation. The generation of reactive oxygen intermediates during the reperfusion phase underlies the pathophysiology of this syndrome. The effects of NO will depend on the context of its production, namely the presence and relative amounts of other reactive intermediates with which NO may interact and the antioxidant status of the tissue. Furthermore it is probably important to differentiate between warm and cold ischemia.

Early evidence for the interaction between NO and $O_2^{\cdot}$ in the liver during ischemia-reperfusion was provided by Bautista and Spitzer who showed that inhibition of NO in the isolated perfused liver increased oxygen radical release [77]. However, Ma and coworkers showed that increased expression of iNOS induced by LPS pretreatment markedly increased liver damage in a NO-dependent manner [78]. Using iNOS knockout mice we have found that the absence of iNOS confers protection from liver damage in both hemorrhagic shock and warm ischemia and reperfusion (unpublished results). It is likely that the combination of NO and $O_2^{\cdot}$ forms peroxynitrite resulting in toxicity in warm ischemia and reperfusion.

Suppression of glutathione levels in this insult probably also contributes to the injury. Cold ischemia and reperfusion may suppress the expression of iNOS or preserve the antioxidant systems and therefore may not exhibit the same NO-dependent injury.

In models of intestinal ischemia-reperfusion, it has been shown NO donors improve survival and decrease intestinal myeloperoxidase activity in the intestine [79], suggesting that NO inhibits neutrophil adherence and migration in the intestinal microcirculation. It was thought that similar effects might be seen in the liver; however, no such benefit was seen in a rodent model of hepatic ischemia-reperfusion [80] or one model of ischemia-reperfusion plus LPS challenge [81].

Hemorrhagic shock represents a common clinical problem faced by clinicians. Decompensated hemorrhagic shock is reached at a volume of blood loss and duration of hypotension after which the vasculature remains hyporesponsive to volume resuscitation or pressors. This syndrome is accompanied by organ dysfunction, including liver injury. In a rodent model of decompensated hemorrhagic shock, infusion of a non-selective inhibitor L-NAME caused an increase in the shock-induced hepatic injury as detected by plasma liver enzymes and liver histology [82]. However, infusion of the iNOS selective inhibitor N6 (imnoethyl)-L-Lysine reduces hepatic damage (unpublished). Thus, the constitutive NOS is essential to preserve perfusion, where as induced NO contributes to hepatic injury.

More direct evidence for a protective role for constitive NO in hepatic ischemia-reperfusion was demonstrated in a rat model of occlusion of the left and medial hepatic lobe vessels for 1 hour, followed by reperfusion for 1 or 24 hours. In this study, rats pretreated with the NOS inhibitor N^G-nitro-L-arginine (L-NNA) developed worsening of endothelial cell and hepatocyte injury, increased release of hepatocyte enzymes, increased lipid peroxidation, and reduction in hepatic blood flow. These effects were abolished by pretreatment with L-arginine [83]. Similarly, a two-hit model of hepatic ischemia-reperfusion followed by administration of LPS results in significant activation of Kupffer cells and subsequent liver injury [84]. In this model, administration of L-NAME worsened the hepatic injury and further impaired microvascular flow; this could be reversed with concomitant infusion of L-arginine or a NO donor. These results are contrasted by our work where iNOS inhibitors reduce hepatic damage in warm ischemia/reperfusion. Similar results were obtained when iNOS knockout mice were compared to wild-type mice (unpublished). Thus, similar to hemorrhagic shock, iNOS expression contributes to damage in isolated ischemia/reperfusion, where as constitutive NO is protective.

The role of nitric oxide in liver tumors

NO has anti-tumor effects as well as known mutagenic effects. NO, reactive oxygen species and, chronic inflammation are associated with carcinogenesis in humans. Infectious agents which result in tissue inflammation and increased NO production such as gastritis, hepatitis and colitis are recognized risk factors for human cancers. In the liver, both viral and helminth infections have been closely associated with the development of tumors. Ohshima *et al* have documented the occurrence of cholangiocarcinoma in the presence of liver fluke infestation [85]. Liu *et al*, in a woodchuck hepatitis virus model of chronic hepatitis, demonstrated the possibility that viral hepatitis increases the risk of liver cancer through a mechanism of increased NO production [86]. In humans, NO levels are elevated in patients with chronic hepatitis and this has been linked with the predisposition to develop liver cancer [87].

The process of carcinogenesis is thought to be a step-wise process involving the inactivation of tumor suppressor genes and the activation of oncogenes by either mutations or deletions of the DNA. Subsequent to DNA damage, cell division must occur for a tumor to develop. NO and its reactive derivatives may play an active role in the multistage process of carcinogenesis by direct cellular cytotoxicity and mutagenicity. The mechanisms of NO-induced cytotoxicity and mutagenicity are numerous and include nitrosative deamination, DNA strand breakage by NO_2^-, oxidative deamination by peroxynitrite, and DNA modification by metabolically activated N-nitrosamines. The formation of N_2O_3 leads to the formation of N-nitroso compounds which deaminate and crosslink DNA. N_2O_3 can also inactivate important enzymes involved in DNA repair mechanisms. As discussed earlier, NO reacts with superoxide anion to form the cytotoxic radical peroxynitrite which is also capable of decomposing to hydroxyl radicals and nitrogen dioxide. The generation of peroxynitrite and other oxidizing agents has also been shown to be genotoxic [88]. Interestingly, DNA deamination results in specific mutations including GC to AT. Sequence data on gene mutations from HBV isolated from chronically infected patients reveal the same GC to AT mutation. This same mutation is evident in the p53 gene from patients with liver cancer. The DNA repair molecule p53 plays an important role in the cellular response to DNA damage from ionizing radiation, UV light, and both exogenous and endogenous chemical mutagens. Forrester *et al.* demonstrated that endogenous NO induces DNA damage, and results in p53 accumulation, which through a negative feedback mechanism inhibits iNOS gene expression. This implies that p53 expression is upregulated in the presence of excessive NO production to prevent NO-induced DNA damage [29]. These data suggest that hepatocytes possess a mechanism to prevent the deleterious effects of NO in the development of liver tumors.

In addition to its carcinogenic potential, it has been shown that NO (from activated macrophages) can kill tumor cells [89]. The target for NO-mediated anti-tumor actions are iron and sulfur containing enzymes involved in mitochondrial respiration. However, whether NO acts as a tumor surveillance mechanism in the liver is unclear. Data from Stadler *et al.* [50], showed that exogenous NO markedly suppressed mitochondrial aconitase as well as complex I and complex II enzymes, two components of the electron transport chain, while endogenously generated NO was less effective. Fukumura and coworkers demonstrated in an *in vivo* and *ex vivo* model using a co-culture system with a hepatoma cell line and Kupffer cells, that induced NO from Kupffer cells produced mitochondrial dysfunction followed by cell membrane disruption leading to tumor cell death [109]. These results suggest that the liver, via Kupffer cell interactions, possesses a mechanism of tumor cell recognition and eradication.

No and cirrhosis

The portal hypertension of cirrhosis is characterized by systemic and splanchnic vasodilation and several mediators have been proposed to have a role including NO [91]. Evidence reveals that LPS induces NO production and consequent vasodilation in cirrhosis. Elevated serum levels of endotoxin are detectable in a large number of cirrhotic patients [92,93], even in the absence of clinical signs of infection. The mechanism is thought to involve the abnormal porto-systemic shunting [94], allowing the bacterial LPS normally cleared by the liver to enter the systemic circulation. This may explain why the hemodynamic effects of cirrhosis are similar to endotoxic sepsis. Elevated NO production in cirrhosis is supported by the finding of increased serum and urine NO_3^- levels [95] and increased urinary cGMP levels [96] in humans with cirrhosis, with a direct correlation between severity of hemodynamic changes and levels of NO_3^-. There is also increased levels of cGMP in aortic segments of cirrhotic rats, which is reduced with chronic L-NAME administration [97]. In animals, glucocorticoids prevent the decrease in SVR in cirrhosis presumably by inhibiting NO production [98]. Further evidence to support this theory was obtained in a rat model of chronic portal hypertension induced by portal vein ligation. In this model of cirrhosis, inhibition of NO synthesis causes significant vasoconstriction in both the systemic and splanchnic beds [99], supporting the role of NO in mediating these derangements. This work was confirmed by studies in carbon tetrachloride-induced cirrhosis in rats, where L-NAME administration corrected the systemic hyperdynamic changes [100]. The vascular production of NO, as measured by aortic cGMP production, was correspondingly reduced in the cirrhotic rats treated with L-NAME. Not all of the NO produced in cirrhosis is detrimental. In animal models of ethanol liver toxicity, inhibition of NOS resulted in increased hepatic microvascular vasoconstriction and hepatocellular damage [101].

Antimicrobial actions of nitric oxide in the liver

Nitric oxide has been shown to be a potent antimicrobial agent against various bacterial, protozoan, parasitic, viral and helminthic infections [102]. The utility of NO as a antimicrobial agent was first recognized by the Sumerians, who used nitrite to cure meats. The commercial food industry later exploited its inhibitory effects on *Clostridium* sporogenesis in the prevention of botulism in canned food products [103]. Since 1987, when Granger *et al.* [104] showed that L-arginine is essential for

Table 2. Pathogens with known NO-mediated anti-microbial effects

Cryptococcus neoformans
Schistosoma mansoni
Trypanosma brucei
Trypanosma cruzi
Toxoplasma gondie
Mycobacterium avium
Leishmania major
Plasmodium vinckei
Plasmodium yoelii
Plasmodium berghei
Plasmodium falciparum
Staphylococcus aureus
Ectromelia virus
Vaccinia virus
Herpes simplex virus type I
Epstein Barr virus

(Reviewed in 109).

the microbiostatic activity of macrophages, it has been shown that NO is associated with extracellular [105,106,107] and intracellular [102,108] mediated cytotoxicity. Table 2 identifies the bacterial, parasitic, and viral agents against which NO has antimicrobial activity [109].

The mechanism by which NO exerts its antimicrobial effects are varied and relate to its small size and lipophilicity. The toxicity of NO against pathogens is typically a result of the elevated and sustained production of NO elicited by cytokine stimulation and by bacterial, fungal, protozoan, and viral antigens. Once formed, NO rapidly diffuses across both prokaryotic cell walls and eukaryotic cell membranes where it is then able to render cytocidal and cytostatic antimicrobial effects. There are several important mechanisms by which NO mediates antimicrobial activity involving the inhibition of cellular replication and energy production. NO interrupts DNA synthesis via inhibition of ribonucleotide reductase [110,111]. Furthermore, NO avidly reacts with intracellular iron-containing molecules and via S-nitrosylation renders them fatally inactive [88,112]. Also, NO in the presence of oxygen can form reactive oxygen intermediates, including peroxynitrite, which have been shown to be directly cytotoxic. To form peroxynitrite, NO reacts with superoxide anion generated by the respiratory burst of activated macrophages. Peroxynitrite then breaks down to produce the highly toxic products: hydroxyl radical and nitrogen dioxide [76]. In addition, NO causes autoribosylation of glyceraldehyde-3-phosphate dehydrogenase, which blocks glycolysis [43]. NO has also been shown to inhibit enzymes important in the Krebs cycle and of the electron transport chain [49,113]. Thus, there are a number of NO-susceptible targets for antimicrobial activity.

The antimicrobial role of NO in the liver is well characterized in malaria, an intracellular pathogen which accounts for over 2 million deaths per year world-wide. Green *et al.* demonstrated that *P. berghei*-infected hepatocytes in culture treated with IFN-γ resulted in a significant decrease in the number of intracellular schizonts [114]. However, the protective effects of IFN-γ are inhibited by L-NMMA [91]. In addition, Mellouk and coworkers reported the potential role of NO in anti-malarial immunity. Here, the typically successful immunization of mice against hepatic stages of *P. berghei* and *P. yeolli* with irradiated sporozoites was blocked by the simultaneous injection of either L-NMMA and anti-IFN-γ antibodies [115]. Thus, NO appears to be an important recognition molecule in the development of anti-malarial immunity. Another mechanism for the anti-malarial effect of NO occurs during the blood-borne stage of the disease by NO diffusing into *Plasmodium.* infected red-blood cells and complexing with cysteine or glutathione to form nitroso-thiol groups which are highly toxic to the merozoites [116].

The anti-malarial effect of NO is also evident in human hepatocytes. First. human hepatocytes have preserved the capacity to produce high. sustained iNOS expression, whereas human monocytes and macrophages are unable to produce comparable levels of NO. This suggests that iNOS expression is well tolerated by human hepatocytes, and since many major infections have a liver component in which NO exerts an important antimicrobial effect, the capacity for iNOS expression has been conserved in the human liver. Whereas hepatocytes my be protected from injury, the same is not true for intracellular pathogens. Mellouk *et al.* demonstrated that IFN-γ leads to elimination of the parasite ***Plasmodium*** *falciparum* in primary human hepatocyte cultures and that the anti-malarial activity is dependent on the cytotoxic molecule NO [115]. Interestingly. the parasite itself stimulates iNOS expression in human hepatocytes independently of added cytokines [117]. In addition, they were able to demonstrate that both spontaneous and IFN-γ-induced inhibition of the malaria parasite could be increased with the addition of the NO synthase cofactors BH_4 and sepiaterin [115]. These results indicate that under *in vitro* conditions, the parasite itself provides a signal that triggers induction of the NO pathway in hepatocytes, and NO formation in infected hepatocytes maybe limited by cofactor availability. This may imply that NO formation may be increased *in vivo* and malaria could be more effectively treated by providing the cofactor for BH_4 to travelers and people living in malaria-endemic countries. Since hepatocytes exhibit potent anti-malarial activity mediated via NO, these results suggest that cells may directly respond to parasite infection by iNOS expression to limit microbial growth.

Conclusions

The regulation of iNOS is tightly controlled and occurs at multiple levels in the gene expression pathway. NO has both beneficial and detrimental effects and understanding the molecular mechanisms that govern iNOS expression is critical to developing novel therapeutic strategies.

References

1. Moncada S, Higgs EA (1991) Endogenous nitric oxide: physiology, pathology and clinical relevance. Eur J Clin Invest 21: 361-374
2. Geller DA, Lowenstein CJ, Shapiro RA, Nussler AK, Di Silvo M, Wang SC, Nakayama DK, Simmons RL, Snyder SH, Billiar TR (1993) Molecular cloning and expression of inducible nitric oxide synthase from human hepatocytes. Proc Natl Acad Sci USA 90: 3491-3495
3. Marsden PA, Schappert KT, Chen HS, Flowers M, Sundell CL, Wilcox JN, Lamas S, Michel T (1992) Molecular cloning and characterization of human endothelial nitric oxide synthase. FEBS Lett 307: 287-293
4. Nakane M, Schmidt HHHW, Pollock JS, Forstermann U, Murad F (1993) Cloned human brain nitric oxide synthase is highly expressed in skeletal muscle. FEBS Lett 316: 175-180
5. Charles IG, Palmer RMJ, Hickery MS, Bayliss MT, Chubb AP, Hall VS, Moss DW, Moncada S (1993) Cloning, characterization, and expression of a cDNA encoding an inducible nitric oxide synthase from the human chondrocyte. Proc Natl Acad Sci USA 90: 11419-11423
6. Sherman PA, Laubach VE, Reep BR, Wood ER (1993) Purification and cDNA sequence of an inducible nitric oxide synthase (NOS) from a human tumor cell line. Biochemistry 32: 11600-11605
7. Hokari A, Zeniya M, Esumi H (1994) Cloning and functional expression of human inducible nitric oxide synthase (NOS) from a glioblastoma cell line A-172. J Biochem 116: 575-581
8. Chartrain N, Geller DA, Koty PP, Sitrin NF, Nussler AK, Hoffman EP, Billiar TR, Hutchinson NI, Mudgett JS. (1994) Molecular cloning, structure, chromosomal localization of the human inducible nitric oxide synthase gene. J Biol Chem 269: 6765-6772
9. Nussler A, Di Silvio M, Billiar TR, Hoffman RA, Geller DA, Selby R, Madriaga J, Simmons RL (1992) Stimulation of the nitric oxide synthase pathway in human hepatocytes by cytokines and endotoxin. J Exp Med 176: 261-264
10. Geller DA, de Vera ME, Russell DA, Shapiro RA, Nussler AK, Simmons RL, Billiar TR (1995) A central role for interleukin-1b in the in vitro and in vivo regulation of hepatic inducible nitric oxide synthase. J Immunol 155: 4890-4898
11. de Vera ME., Shapiro, R.A., Nussler, A.K., Mudgett, J.S., Simmons, R.L., Morris, S.M., Billiar, TR, Geller DA (1996) Transcriptional regulation of human inducible nitric oxide synthase (iNOS) gene by cytokines: Initial analysis of the human iNOS promoter. Proc Natl Acad Sci USA 93: 1054-1059

12. Taylor BS, Kim YM, Wang Q, Shapiro RA, Billiar TR, Geller DA. (1997) Nitric oxide down-regulates hepatocyte-inducible nitric oxide synthase gene expression. Arch Surg 132: 1177-1183
13. de Vera ME, Taylor BS, Wang Q, Shapiro RA, Billiar TR, Geller DA (1997) Dexamethasone Suppresses Inducible Nitric Oxide Synthase Gene Expression by Upregulating I-kBα and Inhibiting NF-κB. Am J Phys (In press)
14. Stuehr DJ, Marletta MA (1985) Mammalian nitrate biosynthesis: mouse macrophages produce nitrite and nitrate in response to Escherichia coli lipopolysaccharide. Proc Natl Acad Sci USA 82: 7738-7742
15. Billiar TR, Curran RD, Stuehr DJ, West MA, Bentz BG, Simmons RL (1989) An L-arginine-dependent mechanism mediates Kupffer cell inhibition of hepatocyte protein synthesis *in vitro*. J Exp Med 169: 1467-1472
16. Curran RD, Billiar TR, Stuehr DJ, Hofmann K, Simmons RL (1989) Hepatocytes produce nitrogen oxides from L-arginine in response to inflammatory products of Kupffer cells. J Exp Med 1989 170: 1769-1774
17. Curran RD, Billiar TR, Stuehr DJ, Ochoà JB, Harbrecht BG, Flint SG, Simmons RL (1990) Multiple cytokines are required to induce hepatocyte nitric oxide production and inhibit total protein synthesis. Ann Surg 212: 462-471
18. Geller DA, Nussler AK, Di Silvio M, Lowenstein CL, Shapiro RA, Wang SC, Simmons RL, Billiar TR. (1993) Cytokines, endotoxin, and glucocorticoids regulate the expression of inducible nitric oxide synthase in hepatocytes. Proc Natl Acad Sci USA 90: 522-526
19. Pittner RA, Spitzer JA (1992) Endotoxin, TNF-α directly stimulate nitric oxide formation in cultured rat hepatocytes from chronically endotoxemic rats. Biochem Biophys Res Commun 185: 430-435
20. Freeswick PD, Wan Y, Geller DA, Nussler AK, Billiar TR (1994) Remote tissue injury primes hepatocytes for nitric oxide synthesis. J Surg Res 57: 205-209
21. Geller DA, Freeswick PD, Nguyen D, Nussler AK, Di Silvio M, Shapiro RA., Wang SC, Simmons RL, Billiar TR. (1994) Differential induction of nitric oxide synthase in hepatocytes during endotoxemia and the acute-phase response. Arch Surg 129: 165-171
22. Knowles RG, Salter M, Brooks SL, Moncada S (1990) Anti-inflammatory glucocorticoids inhibit the induction by endotoxin of nitric oxide synthase. Biochem Biophys Res Commun 172: 1042-1048
23. Geller DA, Di Silvio M, Nussler AK, Wang SC, Shapiro RA, Simmons RL, Billiar TR (1993) Nitric oxide synthase gene expression is induced in hepatocytes in vivo during hepatic inflammation. J Surg Res 55: 427-432
24. Kleinert H, Euchenhofer C, Ihrig-Biedert I, Forstermann U (1996) Glucocorticoids inhibit the induction of nitric oxide synthase II by down-regulating cytokine-induced activity of transcription factor NF-κB. Mol Pharmacol 49: 15-21
25. de Vera ME, Wong J, Zhou JY, Tzeng E, Wong H, Billiar TR, Geller DA (1996) Cytokine-induced nitric oxide synthesis in human liver cells is inhibited by the heat shock response. Surgery 120: 144-149
26. de Vera ME, Kim YM, Wong HR, Wang Q, Billiar TR, Geller DA (1996) Heat shock response inhibits cytokine-inducible nitric oxide synthase expression in rat hepatocytes Hepatology 24: 1238-1245
27. Feinstein DL, Galea E, Aquino DA, Li GC, Xu H, Reis DJ (1996) Heat shock protein 70 suppresses astroglial-inducible nitric oxide synthase expression by decreasing NF-κB activation. J Biol Chem 271: 17724-17732

28. Kim YM, de Vera ME, Watkins SC, Billiar TR (1997) Nitric oxide protects cultured rat hepatocytes from tumor necrosis factor-α-induced apoptosis by inducing heat shock protein 70 expression. J Biol Chem 272: 1402-1411
29. Forrester K, Ambs S, Lupold SE, Kapust RB, Spillare EA, Weinberg WC, Felley-Bosco E, Wang XW, Geller DA, Tzeng E, Billiar TR, Harris CC (1996) Nitric oxide-induced p53 accumulation and regulation of inducible nitric oxide synthase expression by wild-type p53. Proc Natl Acad Sci USA 93: 2442-2447
30. Liu ZZ, Cui S, Billiar TR, Dorko K, Halfter W, Geller DA, Michalopoulos G, Beger HG, Albina J, Nussler AK (1996) Effects of hepatocellular mitogens on cytokine-induced nitric oxide synthesis in human hepatocytes. J leuk Biol 60: 382-388
31. Shaw G, Kamen R (1986) A conserved AU sequence from the 3' untranslated region of GM-CSF mRNA mediates selective mRNA degadation. Cell 46: 659-667
32. Caput D, Beutler B, Hartog K, Thayer R, Brown-Shimer S, Serami A (1986) Identification of a common nucleotide sequence in the 3'-untranslated region of mRNA molecules specifying inflammatory mediators. Proc Natl Acad Sci 83: 1670-1674
33. Han J, Brown T, Beutler B (1990) Endotoxin-reponsive sequences control cachectin/tumor necrosis factor biosynthesis at the translational level. J Exp Med 171: 465-475
34. Pastor CM, Williams D, Yoneyama T, Hatakeyama K, Singleton, Naylor E, Billiar TR (1996) Competition for tetrahydrobiopterin between phenylalanine hydroxylase and nitric oxide synthase in rat liver. J Biol Chem 271: 24534-24538
35. Harada T, Kagamiyama H, Hatakeyama K (1993) Feedback regulation mechanisms for the control of GTP cyclohydrolase I activity. Science 1993 260:1507-1510.
36. West MA, Billiar TR, Mazuski JE, Curran RD, Simmons RL (1988) Endotoxin modulation of hepatocyte secretory and cellular protein synthesis is mediated by Kupffer cells. Arch Surg 123: 1400-1405
37. Billiar TR, Curran RD, West MA, Hofmann K, Simmons RL (1989) Kupffer cell cytotoxicity to hepatocytes in coculture requires L-arginine. Arch Surg 124: 1416-1420
38. Curran RD, Ferrari FK, Kispert PH, Stadler J, Stuehr DJ, Simmons RL, Billiar TR (1991) Nitric oxide and nitric oxide-generating compounds inhibit hepatocyte protein synthesis. FASEB J 5: 2085-2092
39. Billiar TR, Curran RD, Ferrari FK, Williams DL, Simmons RL (1990) Kupffer cell:hepatocyte cocultures release nitric oxide in response to bacterial endotoxin. J Surg Res 48: 349-353
40. Sax HC, Talamini MA, Hasselgren PO, Rosenblum L, Ogle CK, Fischer JE (1988) Increased synthesis of secreted hepatic proteins during abdominal sepsis. J Surg Res 44: 109-116
41. Frederick JA, Hasselgren PO, Davis S, Higashiguchi T, Jacob TD, Fischer JE (1993) Nitric oxide may upregulate *in vivo* hepatic protein synthesis during endotoxemia. Arch Surg 128: 152-157
42. Brass EP, Vetter WH (1993) Inhibition of glucagon-stimulated glycogenolysis by S-nitroso-N-acetylpenicillamine. Pharmacol Toxicol 72: 369-372
43. Horton RA, Leppi ED, Knowles RG, Titheradge MA (1994) Inhibition of hepatic gluconeogenesis by nitric oxide: a comparison with endotoxic shock. Biochem J 299: 735-739
44. Molina y Vedia L, McDonald B, Reep B, Brune B, Di Silvio M, Billiar TR, Lapetina EG (1992) Nitric oxide-induced *S*-nytrosylation of glyceraldehyde-3-phosphate

dehydrogenase inhibits enzymatic activity and increases endogenous ADP-ribosylation. J Biol Chem 267: 24929-24932
45. Ou J, Molina L, Kim YM, Billiar TR (1996) Excessive NO production does not account for the inhibition of hepatic gluconeogenesis in endotoxemia. Am J Physiol 271: G621-G628
46. Nathan C (1992) Nitric oxide as a secretory product of mammalian cells. FASEB J 6: 3051-3064
47. Wood KS, Ignarro LJ (1987) Hepatic cGMP formation is regulated by similar factors that modulate activation of purified hepatic soluble guanylate cyclase. J Biol Chem 262:5 020-5027
48. Billiar TR, Curran RD, Harbrecht BG, Stadler J, Williams DL, Ochoa JB, Di Silvio M, Simmons RL, Murray SA (1992) Association between synthesis and release of cGMP and nitric oxide biosynthesis by hepatocytes. Am J Physiol 262: C1077-C1082
49. Saavedra JE, Billiar TR, Williams DL, Kim YM, Watkins SC, Keefer LK (1997) Targeting nitric oxide (NO) delivery in vivo. Design of a liver-selective NO donor that blocks tumor necrosis factor-a induced apoptosis and toxicity in the liver. J. Med. Chem. 40: 1947-1954
50. Stadler J, Billiar TR, Curran RD, Steuhr DJ, Ochoa JB, Simmons RL (1991) Effects of exogenous and endogenous nitric oxide on mitochondrial respiration of rat hepatocytes. Am J Physiol 260. C910-C916
51. Kurose I, Kato S, Ishii H, Fukumura D, Miura S, Suematsu M, Tsuchiya M (1993) Nitric oxide mediates lipopolysaccharide-induced alteration of mitochondrial function in cultured hepatocytes and isolated perfused liver. Hepatology 18: 380-388
52. Ghezzi P, Saccardo B, Villa P, Rossi V, Bianchi M, Dinarello CS (1986) Role of interleukin-1 in the depression of liver drug metabolism by endotoxin. Infect Immun 54: 837-840
53. Peterson TC, Renton KW (1986)The role of lymphocytes, macrophages, and interferon in the depression of drug metabolism by dextran sulfate. Immunopharmacology 11: 21-28
54. Khatsenko OG, Gross SS, Rifkind AB, Vane JR (1993) Nitric oxide is a mediator of the decrease in cytochrome P-450-dependent metabolism caused by immunostimulants. Proc Natl Acad Sci USA 90: 11147-11151
55. Stadler J, Trockfeld J, Schmalix WA, Brill T, Siewert JR, Greim H, Doehmer J (1994) Inhibition of cytochromes P4501A by nitric oxide. Proc Natl Acad Sci USA 91. 3559-3563
56. Ochoa JB, Udekwu AO, Billiar TR, Curran RD, Cerra FB, Simmons RL, Peitzman AB (1991) Nitrogen oxide levels in patients after trauma and during sepsis. Ann Surg 214: 621-626
57. Billiar TR, Curran RD, Harbrecht BG, Stuehr DJ, Demetris AJ, Simmons RL (1990) Modulation of nitric oxide synthesis in vivo: N^G-monomethyl-L-arginine inhibits endotoxin-induced nitrite/nitrate synthesis while promoting hepatic damage. J Leuk.Biol 48: 565-569
58. Harbrecht BG, Billiar TR, Stadler J, Demetris AJ, Ochoa J, Curran RD, Simmons RL (1992) Inhibition of nitric oxide synthesis during endotoxemia promotes intrahepatic thrombosis and an oxygen radical-mediated hepatic injury. J Leuk Biol 52: 390-394
59. Harbrecht BG, Stadler J, Demetris AJ, Simmons RL, Billiar TR (1994) Nitric oxide and prostacyclin interact to prevent hepatic damage during murine endotoxemia. Am J Physiol 266: G1004-G1010

60. Harbrecht BG, Billiar TR, Stadler J, Demetris AJ, Ochoa JB, Curran RD, Simmons RL (1992) Nitric oxide synthesis serves to reduce hepatic damage during acute murine endotoxemia. Crit Care Med 20: 1568-1574
61. Szabo C, Southan GJ, Thiemermann C (1994) Beneficial effects and improved survival in rodent models of septic shock with S-methylisothiourea sulfate, a potent and selective inhibitor of inducible nitric oxide synthase. Proc Natl Acad Sci USA 91: 12472-12476
62. Ou J, Carlos TM, Saavedra JE, Keefer LK, Watkins SC, Harbrecht BG, Billiar TR (1994) Evidence that NOS3 protects against hepatic necrosis while NOS2 prevents apoptosis in rodent endotoxemia. (In press).
63. Bursch W, Oberhammer F, Jirtle RL, Askari M, Sedivy R, Grasl-Kraupp B, Purchio AF, Schulte-Hermann R (1993) Transforming growth factor-β1 as a signal for induction of cell death by apoptosis. Br J Cancer 67: 531-536
64. Watanabe-Fukunaga R, Brannan CI, Copeland NG, Jenkins NA, Nagata S (1992) Lymphoproliferation disorder in mice explained by defects in Fas antigen that mediates apoptosis. Nature 356: 314-317
65. Leist M, Gantner F, Jilg S, Wendel A (1995) Activation of the 55 kDa TNF receptor is necessary and sufficient for TNF-induced liver failure, hepatocyte apoptosis, and nitrite release. J Immunol 154: 1307-1316
66. Albina JE, Cui S, Mateo RB, Reichner SJ (1993) Nitric oxide mediated apoptosis in murine peritoneal macrophages. J Immunol 150: 5080-5085
67. Lipton SA, Choi YB, Pan ZH, Lei SZ, Chen HS, Sucher NJ, Loscalzo J, Singel DJ, Stamler JS (1993) A redox-based mechanism for the neuroprotective and neurodestructive effects of nitric oxide and related nitroso compounds. Nature 364: 626-632
68. Martin-Sanz P, Diaz-Guerra MJM, Casado M, Bosca L (1996) Bacterial lipopolysaccharide antagonizes transforming growth factor-beta 1-induced apoptosis in primary cultures of hepatocytes. Hepatology 23: 1200-1207
69. Bohlinger I, Leist M, Barsig J, Uhlig S, Tiegs G, Wendel A (1995) Interleukin-1 and nitric oxide protect against tumor necrosis α-induced liver injury through distinct pathways. Hepatology 22: 1829-1837
70. Kim YM, Talanian RV, Billiar TR. (1997) Nitric oxide regulates apoptosis by redox-based inhibition of caspase-3-like protease. J. Biol. Chem. (In press).
71. Kuo PC, Slivka A (1994) Nitric oxide decreases oxidant-mediated hepatocyte injury. J Surg Res 56: 594-600
72. Arthur MJP (1988) Reactive oxygen intermediates and liver injury. J Hepatol 6: 125-131
73. Kanner J, Harel S, Granit R (1991) Nitric oxide as an antioxidant. Arch Biochem Biophys 289: 130-136
74. Arthur MJP, Bentley IS, Tanner AR, Kowalski Saunders P, Millward-Sadler GH, Wright R (1985) Oxygen-derived free radicals promote injury in the rat. Gastroenterology 89: 1114-1122
75. Kim YM, Bergonia H, Lancaster JR (1995) Nitrogen oxide-induced autoprotection in isolated rat hepatocytes. FEBS Letters 374: 228-232
76. Beckman JS, Beckman TW, Chen J, Marshall PA, Freeman BA (1990) Apparent hydroxyl radical production by peroxynitrite: implications for endothelial injury from nitric oxide and superoxide. Proc Natl Acad Sci USA 87: 1620-1624
77. Bautista AP, Spitzer JJ (1994) Inhibition of nitric oxide formation *in vivo* enhances superoxide release by the perfused liver. Am J Physiol 266: G783-G788

78. Ma TT, Ischiropoulos H, Brass CA (1995) Endotoxin-stimulated nitric oxide production increases injury and reduces rat liver chemiluminescence during reperfusion. Gastroenterology 108: 463-469
79. Christopher TA, Ma X, Lefer AM (1994) Beneficial actions of *S*-nitroso-*N*-acetylpenicillamine, a nitric oxide donor, in murine traumatic shock. Shock 1: 19-24
80. Jaeschke H, Schini VB, Farhood A (1992) Role of nitric oxide in the oxidant stress during ischemia/reperfusion injury of the liver. Life Sci 50: 1797-1804
81. Liu P, Yin K, Yue G, Wong PYK (1996) Role of nitric oxide in hepatic ischemia-reperfusion with endotoxemia. J Inflamm 46: 144-154
82. Harbrecht BG, Wu B, Watkins SC, Marshall HP Jr, Peitzman AB, Billiar TR (1995) Inhibition of nitric oxide synthase during hemorrhagic shock increases hepatic injury. Shock 4: 332-337
83. Kobayashi H, Nonami T, Kurokawa T, Takeuchi Y, Harada A, Nakao A, Takagi H (1995) Role of endogenous nitric oxide in ischemia-reperfusion injury in rat liver. J Surg Res 59: 772-779
84. Wang Y, Mathews WR, Guido DM, Farhòod A, Jaeschke H (1995) Inhibition of nitric oxide synthesis aggravates reperfusion injury after hepatic ischemia and endotoxemia. Shock 4: 282-288
85. Ohshima H, Brouet IM, Bandaletova H, Adachi S, Oguchi S, Iida S, Kurashima Y, Morishita Y, Sugimura T, Esumi H (1992) Polyclonal antibody against an inducible form of nitric oxide synthase purified from the liver of rats treated with *Propionibacterium acnes* and lipopolysaccharide. Biochem Biophys Res Commun 187: 1291-1297
86. Liu RH, Jacob JR, Hotchkiss JH, Cote PJ, Gerin JL, Tennant BC (1994) Woodchuck hepatitis virus surface antigen induces nitric oxide synthesis in hepatocytes: possible role in carcinogenesis. Carcinogenesis 15:2875-2877.
87. Ohshima H, Bartsch H (1994) Chronic infections and inflammatory processes as cancer risk factors: possible role of nitric oxide in carcinogenesis. Mutat Res 305:253-264.
88. Keefer LK, Wink DA (1996) DNA damage and nitric oxide. Adv Exp Med Biol 387: 177-185
89. Stuehr DJ, Nathan CT (1989) Nitric oxide. A macrophage product responsible for cytostasis and respiratory inhibition in tumor target cells. J Exp Med 169: 1543-1555
90. Fukumura D, Yonei Y, Kurose I, Saito H, Ohishi T, Higuchi H, Miura S, Kato S. Kimura H, Ebinuma H, Ishi H (1996) Role of nitric oxide in Kupffer cell-mediated hepatoma cell cytotoxicity *in vitro* and *ex vivo*. Hepatology 24: 141-149
91. Korthuis RJ, Benoit JN, Kvietys PR, Townsley MI, Taylor AE, Granger DN (1985) Humoral factors may mediate increased rat hindquarter blood flow in portal hypertension. Am J Physiol 249: H827-H833
92. Lumsden AB, Henderson JM, Kutner MH (1988) Endotoxin levels measured by a chromogenic assay in portal, hepatic and peripheral venous blood in patients with cirrhosis. Hepatology 8: 232-236
93. Triger DR, Boyer TD, Levin J (1978) Portal and systemic bacteremia and endotoxaemia in liver disease. Gut 19: 935-939
94. Groszmann R, Kotelanski B, Cohn JN, Khatri IM (1972) Quantitation of portosystemic shunting from the splenic and mesenteric beds in alcoholic liver disease. Am J Med 53: 715-722

95. Hori N, Okanoue T, Mori T, Kashima K, Nishimura M, Nanbu A, Yoshimura M, Takahashi H (1996) Endogenous nitric oxide production is augmented as the severity advances in patients with liver cirrhosis. Clin Exp Pharmacol Physiol 23:30-35.
96. Miyase S, Fujiyama S, Chikazawa H, Sato T (1990) Atrial natriuretic peptide in liver cirrhosis with mild ascites. Gastroenterol Jpn 25: 356-362
97. Niederberger M, Gines P, Tsai P, Martin PY, Morris K, Weigert A, McMurtry I, Schrier RW (1995) Increased aortic cyclic guanosine monophosphate concentration in experimental cirrhosis in rats: evidence for a role of nitric oxide in the pathogenesis of arterial vasodilation in cirrhosis. Hepatology 21: 1625-1631
98. Radomski MW, Palmer RMJ, Moncada S (1990) Glucocorticoids inhibit the expression of an inducible, but not constitutive, nitric oxide synthase in vascular endothelial cells. Proc Natl Acad Sci USA 87: 10043-10047
99. Pizcueta P, Pique JM, Bosch J, Whittle BJR, Moncada S (1991) Effects of endogenous nitric oxide inhibition on the hemodynamic changes of portal hypertensive rats (abstr). Gastroenterology 100: A785
100. Niederberger M, Martin PY, Gines P, Morris K, Tsai P, Xu DL, McMurtry I, Schrier RW (1995) Normalization of nitric oxide production corrects arterial vasodilation and hyperdynamic circulation in cirrhotic rats. Gastroenterology 109: 1624-1630
101. Oshita M, Takei Y, Kawano S, Hijioka T, Masuda E, Goto M, Nishimura Y, Nagai H, Iio S, Tsuji S, Fusamoto H, Kamada T (1994) Endogenous nitric oxide attenuates ethanol-induced perturbation of hepatic circulation in the isolated perfused rat liver. Hepatology 20: 961-965
102. Vouldoukis I, Riveros-Moreno V, Dugas B, Ouaaz F, Becherel P, Debre P, Moncada. S, Mossalayai MD (1995) The killing of *Leishmania major* by human macrophages is mediated by nitric oxide induced after ligation of Fc epsilon RII/CD23 surface antigen. Proc Natl Acad Sci USA 92: 7804-7808
103. Woods LF, Wood JM, Gibbs PA (1981) The involvement of nitric oxide in the inhibition of the phosphorclastic system in *Clostridium sporogenes* by sodium nitrite. J Gen Microbiol 125: 399-406
104. Granger DL, Hibbs JB Jr., Perfect JR, Durack DT (1988) Specific amino acid (L-arginine) requirement for the microbiostatic activity of murine macrophages. J Clin Invest 81: 1129-1136
105. James SL, Glaven J (1989) Macrophage cytotoxicity against schistosomula of *Schistosoma mansoni* involves arginine-dependent production of reactive nitrogen intermediates. J Immunol 143: 4208-4212
106. Vincendeau P, Daulouede S, Veyret B, Darde ML, Bouteille B, Lemesre JL (1992) Nitric oxide mediated cytostatic activity on *Trypanosoma brucei gambiense* and *Trypanosoma brucei brucei*. Exp Parasitol 75: 353-360
107. Munoz-Fernadez, MA., Fernandz, M., Fresno, M (1992) Activation of human macrophages for the killing of intracellular *Trypanosoma cruzi* by TNF-α and IFN-γ through the nitric oxide-dependent mechanism. Immunol Lett 33: 35-40
108. Flesch IE, Hess JH, Kaufmann SH (1994) NADPH diaphorase staining suggests a transient and localized contribution of nitric oxide to host defence against an intracellular pathogen *in situ*. Int Immunol 6: 1751-1757
109. Vouldoukis I, Mazier D, Debre P, Mossialayi MD (1995) Nitric oxide and human infectious disease. Res Immunol 146: 689-692

110. Lepoivre M, Chenais B, Yapo A, Lemaire G, Thelander L, Tenu JP (1990) Alterations of ribonucleotide reductase activity following induction of the nitrite-generating pathway in adenocarcinoma cells. J Biol Chem 265: 14143-14149
111. Kwon NS, Stuehr DJ, Nathan CF (1991) Inhibition of tumor cell ribonuclease reductase by macrophage-derived nitric oxide. J Exp Med 174: 761-767
112. Hibbs JB Jr., Taintor RR, Vavrin Z, Rachlin EM (1988) Nitric oxide: a cytotoxic activated macrophage effector molecule. Biochem Biophys Res Comm 157: 87-94
113. Spence JT, Merrill MJ, Pitot HC (1981) Role of insulin, glucose, and cyclic GMP in the regulation of glucokinase in cultured hepatocytes. J Biol Chem 256: 1598-1603
114. Green SJ, Mellouk S, Hoffman SL, Meltzer MS, Nacy CA (1990) Cellular mechanisms of non-specific immunity to intracellular infection: cytokine-induced synthesis of toxic nitrogen oxides from L-arginine by macrophages and hepatocytes. Immunnol Lett 25: 15-19
115. Mellouk S, Hoffman SL, Liu ZZ, de la Vega P, Billiar TR, Nussler AK (1994) Nitric oxide-mediated antiplasmodial activity in human and murine hepatocytes induced by gamma interferon and the parasite itself: enhancement by exogenous tetrahydrobiopterin. Infect Immun 62: 4043-4047
116. Rockett KA, Awburn MM, Aggarwal BB, Cowden WB, Clark IA (1992) *In vivo* induction of nitrite and nitrate by tumor necrosis factor, lymphotoxin, and interleukin-1: possible roles in malaria. Infect Immun 60: 3725-3730
117. Nussler A, Renia L, Pasquetto V, Miltgen F, Matile H, Mazier D (1993) *In vivo* induction of the nitric oxide pathway in hepatocytes after injection with irradiated malaria sporozoites, malaria blood parasites or adjuvants. Eur J Immunol 23: 882-887

Index

MIX
Papier aus verantwortungsvollen Quellen
Paper from responsible sources
FSC® C105338

If you have any concerns about our products,
you can contact us on
ProductSafety@springernature.com

In case Publisher is established outside the EU,
the EU authorized representative is:
Springer Nature Customer Service Center GmbH
Europaplatz 3, 69115 Heidelberg, Germany

Printed by Libri Plureos GmbH
in Hamburg, Germany